SWEDENBORG'S 1714 AIRPLANE

A MACHINE TO FLY IN THE AIR

SWEDENBORG'S 1714 AIRPLANE

A MACHINE TO FLY IN THE AIR

BY HENRY SÖDERBERG

GEORGE F. DOLE, EDITOR

SWEDENBORG FOUNDATION
NEW YORK

First Printing 1988
ISBN 0-87785-138-7
Library of Congress Catalog Card Number 88-60067

Cover and Book Design
by Renee Clark, Superior Graphics

Cover Art, picture in oil of Emanuel Swedenborg
drawing a sketch of his "Machine to Fly in the Air",
painted by Otto Nielsen exclusively for
Scandinavian Airlines System, (SAS) in 1960

Disk Conversion, Typography, and Pagination
by Brad Allan Walrod, High Text Graphics

Swedenborg Foundation, Inc.
139 East 23rd Street
New York, NY 10010

Manufactured in the United States of America

In celebration of the 300th anniversary of the birth of Emanuel Swedenborg, the Swedenborg Foundation has been privileged to have the author, Mr. Söderberg, present in special ceremonies, advance copies of this book to ~

His Majesty the King of Sweden
Carl XVI Gustaf
and to the Swedenborg Family.

Young Emanuel Swedenborg's proposal for "A Machine to Fly in the Air" has been recognized as the first rational design for heavier-than-air flight.

This book, based on several years of research in libraries, archives, and aviation institutions around the world, represents the first comprehensive study of this remarkable design.

This publication has been produced through the joint efforts of the Publisher and Scandinavian Airlines System, (SAS).

The Swedenborg Foundation, New York
Publisher

ACKNOWLEDGMENTS

Although I have done my own research and planning for this book, I owe a great deal to a number of libraries, institutions, and individuals for their generous advice and assistance during the four years that this work was in the making.

The book would never have come about without the initiative, continuous encouragement, and active participation of the Swedenborg Foundation and its staff.

It would not have been possible to carry out my extensive research, involving repeated visits to universities and libraries in many countries, without the direct and indirect support of the Scandinavian Airlines System (SAS) and its Public Relations Department in Stockholm. I am especially indebted to *Gunnel Thörne*, in charge of research and archives.

Wherever I have gone, I have found the personnel without exception to be extremely understanding and helpful, and sometimes unexpectedly enthusiastic about the subject matter itself. Many have helped and served beyond the call of duty. The principal institutions visited have been the following:

> University libraries in Stockholm, Oslo, Gothenburg, and Upsala; New York and Columbia Universities, New York; Marshall Law School, Chicago; MIT, Cambridge; University of California, Los Angeles; Southern Methodist University, Dallas; University of North Dakota, Grand Forks; McGill University, Montreal; and the University of Cologne, Federal Republic of Germany.

Some of the libraries have been more specialized than others in matters related to Swedenborg. They have been my primary sources when collecting information about the flying machine, and I can warmly recommend them to future students of Swedenborg-related matters. I would express my gratitude particularly to the following individuals who have given exceptionally good help and advice:

The Rev. *Olle Hjern* of the Swedenborg Society, Stockholm; *David Glenn*, Librarian of the Swedenborg Academy, Bryn Athyn, PA; Mrs. *Nancy Lee* of the Swedenborg Academy, Glenview, IL; *Tom Crouch*, Curator, and *Lawrence Wilson*, Research Technician, of the Smithsonian National Air and Space Museum, Washington, D.C.; Dr. *Leonard Bruno*, Science Specialist, The Library of Congress, Washington, D.C.; *A. W. L. Nayler*, Librarian, The Royal Aeronautical Society, London; *Duane Reed*, Librarian for Special Collections of the Library of the United States Air Force Academy, Colorado Springs, CO; *Kajsa Tegnér* and *Ante Strand*, Librarians for Rarities of the Library of the Royal Academy of Science, University of Stockholm; *Allan Ranius*, Curator, and *Mathias von Wachenfeld* Librarian, of the Diocesan Library of Linköping; and *Åke Norrgård*, Library Assistant of the Technical Museum, Stockholm.

I have also enjoyed the generous and friendly assistance of the personnel of the libraries of the following aviation organizations: The International Civil Aviation Organization (ICAO), Montreal; The International Air Transport Association (IATA), Geneva and Montreal; L'Institut du Transport Aérien (ITA), Paris; also the Royal Library (KB) and the City Library of Stockholm, the Aeronautical Research Institute (FAO) of Sweden, Stockholm; the New York City Public Library; and the Library of the Deutsches Museum von Meisterwerken der Naturwissenschaft und Technik, Munich.

In response to inquiries disseminated through *Logos*, the Newsletter of the Swedenborg Foundation, I have received many letters from interested people in the United States, England, Australia, Japan, and elsewhere. Such letters have regularly served as sources of inspiration, and some have made constructive contributions to my research. I have not been able to respond personally to them all, and in cases where I have not, I should like to express here my thanks for the writers' interest and notable enthusiasm.

I owe a particular debt of thanks to my friends of many years, *Hans Erik Löfkvist*, retired Executive Vice President of SAAB Aeroplane Manufacturers, Linköping, and *Birger Holmer*, retired Vice President, Aircraft Research and Development, Scandinavian Airlines System, for their analyses of and comments on the technical aspects of the Swedenborg airplane. Their contributions, based on longtime professional knowledge and experience, have added an extra dimension of credibility to this book.

Last but not least, I owe a great debt to *Dr. George Dole*, Chairman of the Editorial and Publication Committee of the Swedenborg Foundation, for his editing of my original manuscript into a more natural and flowing English than this particular Scandinavian could produce.

CONTENTS

VIII. MODELS OF "THE MACHINE"

APPENDIX:
SWEDENBORG'S CONTACTS ABROAD
RELATED TO HIS MECHANICAL INTERESTS

NOTES

BIBLIOGRAPHY

INDEX

PREFACE

Writing a book exclusively about Swedenborg's suggestion for a flying machine might at first glance seem disproportionate. Why give so much attention to a craft that never flew and that had no apparent influence on subsequent inventors and experimenters, especially since Emanuel Swedenborg's reputation in the fields of science, philosophy, and religion is certainly secure without it?

Five years ago, I would have argued in much this way. In the course of research into Scandinavian contributions to aviation, however, it became apparent—especially in view of the availability of new evidence and the growing interest in the historical roots of aviation—that a study of Swedenborg's suggestion in a wider context was both appropriate and timely. The primary reason for this is the incontestable fact that his was the first rational design for heavier-than-air flight. Aviation historians and Swedenborg scholars are unanimous in this respect.

We should not be put off, as many have been, by the appearance of the craft. Admittedly his sketch (which has been realized in several exhibition models) may raise smiles when compared to modern airplanes—a bug, a turtle, a saucer, an insect— have its looks led people to overlook the remarkable insights it incorporates?

Let us pause for a moment and remember that Swedenborg was some 150 years ahead of his time with this design. In the beginning of the eighteenth century, very little was known about the laws of physics. Newton's ideas were just beginning to spread, and had hardly reached Sweden. The only sources of energy known were those of man, beast, wind, fire, and water. Further, flying itself was seen at best as an utter impossibility, at worst as a matter of witchcraft, in defiance of God's laws. Those who dabbled in it risked their reputations.

When a model of the Swedenborg airplane was presented to the Smithsonian National Air and Space Museum in 1961, the Curator of the museum, Paul E. Garber, had the following to say:

We are inclined to smile as we compare this design of 250 years ago with the aircraft in which we fly today. As we hear the powerful roar of airplane engines at a speed of six hundred miles an hour, we are amused by the little figure of a man trying to row his way through the air with paddles. But it is not from the viewpoint of 1961 that we should consider the concept of Swedenborg. Rather we should look at it through the eyes and mind of the year 1714 when it was conceived.

It is particularly fitting that this presentation took place in the board room of Scandinavian Airlines System (SAS) in New York, and that today the model is one of the Smithsonian's treasures, placed in the very special historical chamber called the Early Flights Room.

What really caught my attention and triggered off the determination to look deeper into the matter was an unexpected discovery in 1982 in the library of the Massachusetts Institute of Technology in Cambridge. I found there the July, 1910 issue of England's prestigious *Journal of the Royal Aeronautical Society*. In it I read not only the judgment that Swedenborg's airplane was the "first rational proposal for a flying machine," but also that a manuscript embodying this invention, with a drawing, was preserved in the Diocesan Library of the city of Linköping, Sweden—where I was born, and where I lived and went to school for many years, where I still have close ties with relatives and friends, which was the proving ground for Sweden's flying pioneers at the beginning of this century, and which is today, as the site of the SAAB factories, the main center for airplane production in northern Europe. It was more than a little embarrassing to discover that this treasure had been in my own back yard, so to speak, and that no teacher, priest, or librarian had ever mentioned it.

When I next returned to Sweden, I soon discovered that there was little awareness of the Swedenborg airplane. Even learned and intelligent friends of mine, experts in aeronautics, assured me, when I told them about my discovery, that it was Leonardo da Vinci who had invented the airplane—which is, of course, quite wrong. One even suggested that it was Michelangelo. Some had no idea whatsoever where Swedenborg came into the picture, others had very vague ideas about his strange "saucer." This, unfortunately, seems symptomatic of the attitude of many of us Swedes toward our eminent compatriot; he is better known and more highly respected abroad than at home.

Fortunately, circumstances favored my growing interest in a thorough study of this subject. I had committed myself, on behalf of Scandinavian Airlines System, to a broad investigation of the contributions of Scandinavians to international aviation when I retired in 1981, after thirty-five years of active work for the Swedish air authorities and for SAS. Since this undertaking involved visiting virtually all the major aviation libraries and archives in the Western world, I had an unparalleled opportunity to discover what had been written about Swedenborg's airplane, soon discovering that the references were generally disjointed, showing little awareness of each other or of other dimensions of Swedenborg's life and thought.

Particularly for this latter information, it was necessary to turn to the resources of Swedenborgian churches and libraries, where in many cases there are excellent collections. After all, everything Swedenborg did, including his airplane design, needs to be seen in the larger context of his life, his work, and his character. It would be misleading to describe the airplane without investigating his sources of inspiration, his early travels, and his relation with the eminent Swedish inventor Christopher Polhem and with Charles XII.

An especially stimulating facet of my research has been the examination of the various models of the craft which have been built from time to time by devoted Swedenborgians. These have helped me understand better the uniqueness and brilliance of Swedenborg's early inventions, some of them so advanced that they were not understood by his contemporaries.

Some of my Swedish friends, experts of high standing in the fields of aerodynamics and airplane engineering, have provided me with up-to-date evaluations of the design from a purely technical point of view. Their analyses, together with those of earlier experts (including those of the Smithsonian Air and Space Museum), serve to reinforce the judgment that while Swedenborg cannot be said to have "invented the airplane," he must be recognized as having proposed the first rational design for heavier-than-air flight.

From all the sources mentioned, and by making inquiries of Swedenborgians and friends of Swedenborg around the world through the Swedenborg Foundation newsletter *Logos*, I have to the best of my ability exhausted the available sources of information, and the present work is as thoroughly documented as possible. There may of course be material which has not come to light. In particular, there was probably additional information in a diary and notes which Swedenborg shipped to Sweden from Hamburg in 1714, but which never arrived. The history of scholarship leaves open the possibility that it may yet be found in some forgotten corner.

The present study, then, is not intended to "close the book." Aviation has now reached a stage of maturity which includes a natural curiosity about its roots, its "firsts." The research in aviation history will no doubt expand and become more intense over the coming years, with welcome new discoveries correcting and clarifying our present understanding.

There is one small technicality to note before proceeding. The individual we know as Emanuel Swedenborg was named Emanuel Swedberg until 1719, when the family was ennobled and the name Swedenborg taken. While it would thus be pedantically more accurate to speak of the airplane designer as Swedberg, I follow common practice in using the name by which he has come to be most widely known.

New York, October 1987

SWEDENBORG'S 1714 AIRPLANE

A MACHINE TO FLY IN THE AIR

DREAMS OF FLIGHT

WHO INVENTED THE AIRPLANE?

Who invented the airplane?

Well-informed people usually name either Leonardo da Vinci or the American brothers Wilbur and Orville Wright.[1]

Both answers are wrong.

Wilbur Wright himself, some seven years before the first flight, set the question in its true perspective. "I believe," he said, "that simple flight, at least, is possible to man and that the experiments and investigations of a large number of independent workers will result in the accumulation of information and knowledge and skill which will finally lead to accomplished flight. . . . I wish to avail myself of all that is already known and then, if possible, add my mite to help on the future workers who will attain final success."

What then about Leonardo da Vinci?

As far as we know, Leonardo (1452-1519)—artist, musician, architect, and man of science—was the first to study the problem of flying in a theoretical and scientific manner. He, like many of his successors, studied how birds flew, and made designs and drawings to illustrate his findings. The results of his research were presented in a book published in 1504, *Codice sur Volo Degli Uccelli*. He approached the problem on the apparent assumption that human flight could be achieved by imitating birds mechanically; and the bulk of his notes therefore dealt with a combination of bird and human aviation.[2]

To this end he designed an ornithopter[3] whose wings were to be attached to spars which the pilot could operate by means of his hands and by foot pedals. In the *prone type*, the pilot was to lie on a wooden frame and operate either two or four wings by means of a complex system of foot stirrups and levers; in the *standing type*, the pilot would be upright and work an equally complicated (and exceptionally heavy) mechanism to activate the wings.

Leonardo's ornithopter was unworkable, and so were all the other ornithopter designs and models produced by scores of dreamers, experimenters, and inventors for centuries to come. In fact, the noted Italian scholar G. A. Borelli demonstrated in his 1680 work *De Motu Animalium* that it would be impossible for a human being to fly using only his own strength.[4] He converted only a few: many others continued stubbornly along the ornithopter road—to failure.[5]

Despite the impracticability of his ornithopter, Leonardo does have a place in the history of aviation. The first experimentalist in a scientific sense, giving hints of Newton's law of gravitation, he both described and designed two genuine forerunners, the parachute and the helicopter. The latter was particularly remarkable in its originality—a screw with which one could climb and descend in the air.

While we cannot credit him with the invention of the airplane, then, we must rank him high among the pioneers, truly outstanding in his time. Regrettably, his designs had no direct influence on the achievement of flight. After his death, many of his manuscripts and drawings remained unknown to the outside world. Recognition came only when Napoleon looted the bulk of the material from Italy in 1796, and when in 1797 the French scholar B. Venturi published the sensational *Essai sur les ouvrages Physico-Mathématiques de Leonardo da Vinci*. Even this, however, did not include Leonardo's inventions and drawings in the mechanical field. It was not until 1874 that an article in a French aviation journal brought these secrets to light.[6]

Such was the sad fate of Leonardo da Vinci's unique and advanced thought in the field of aeronautics. The outstanding British aviation historian Charles Gibbs-Smith writes:

> His work, if it had been properly known to his contemporaries, must have changed the whole course of science and technology: his aeronautical work, whatever its limitations, would immediately have stimulated others to test and correct his theories.... he has contributed to the general inspiration of those who have known his work for the last 150 years. Such inspiration has often in itself been a positive and powerful force in aeronautical history. There is at least one great figure in modern aviation who can trace his life's work direct to Leonardo... Igor Sikorsky... his life work has been the helicopter.[7]

THOUSANDS OF YEARS, MEN, AND IDEAS...

The first securely documented ascent into the air took place on November 21, 1783, when two Frenchmen, the pilot de Rozier and the Marquis d'Arlandes went up in a hot air balloon fabricated by the Montgolfier brothers.[8] We should note, however, that it would require another hundred and twenty years to solve the problems of heavier-than-air flight.

The story that leads up to this first ascent is a long and fascinating one, covering century after century of dreaming and experimenting. It is full of imaginative and bold men, inventors, dreamers, visionaries, religious fanatics, and mystics, many of them willing to sacrifice arms, legs, and lives in order to prove that human beings could conquer the sky. They dived from towers and walls, they flapped their home-made wings of canvas, fabrics, and leather, they floundered down to break their bones.[9] Then there were also the learned men, either less daring or more humble, who stayed on terra firma and drew sketches of their aerial fantasies.

The thousands of men and thousands of years produced thousands of different ideas. Models were taken both from nature and from the realm of imagination—birds and flying fish, butterflies, salamanders, bats, demons, cherubs, angels, anything on earth or in heaven that could fly. Conceptions of flight found their way into poetry, legend, philosophy, religion, and only eventually into natural science.

SOME NAMES FROM THE PREHISTORY OF FLIGHT

There are some hundreds of books that recapitulate the story of human efforts to fly.[10] Most begin with early experiments in such places as Greece, China, Mexico, Egypt, Assyria, and Babylonia. The dream of flight seems to have been an insistent theme in ancient cultures, crossing and recrossing the border between actual attempts on the one hand, and myth and legend on the other. If messengers could fly down from heaven and then return, perhaps mortals could learn to fly up. Indeed, how many children in the Western world have modeled their private images of flight on their ideas of angels?

This is not the place to retell the long story in detail. It is appropriate, however, to take note of some of the more significant figures, in order to see Swedenborg's achievement in perspective. We should be mindful in particular of the fact that at the time of his design, neither the steam engine nor hydrogen gas was known.[11]

In his excellent book, *The Prehistory of Flight*,[12] Clive Hart provides a thorough treatment of the subject in its historical, philosophical, literary, and religious aspects. From his list of some fifty names we may select only a few of those who predated the first successful ascent in 1783.[13]

850 B. C. King Bladud of Britain, according to legend, tried with wings attached to his arms to jump from a tower at Troja Nova (London), but fell on the temple of Apollo and killed himself.

400 B. C. Archytas of Tarentum made a mechanical dove which was supposed to move "by a current of air hidden and enclosed with it"; most likely this was a kind of flying object similar to a kite, produced only in model form.

1010 A. D. Eilmer, a monk at Malmesbury, England, attached wings to his hands and feet, jumped down from the top of a tower, and broke his legs.

1250 A friend of Roger Bacon[14] constructed a flying boat equipped with wings which could be moved with a crank handle; there is no record of any practical trials.

1498 Giovanni Battista Danti, in Perugia, Italy, constucted a flying device with feather wings attached to a structure of iron bars. Trial flights were made from a tower at the city square; the vehicle crashed on the roof of the nearby St. Mary's Church.

1500-1505 Leonardo da Vinci made his famous studies of aeronautics. In 1505, he seems to have tried out one of his interesting and highly complex ornithopters, without success.[15]

1507 John Damian, an Italian adventurer, tried to fly with wings made of hen feathers from the walls of Stirling Castle in Scotland—a very short trip.

1589 John Williams, at Conway Castle in Wales, tried to use a long coat as a sail or as wings; he fell immediately onto a stone and was emasculated. He was seven years old at the time.

1647 Tito Livio Burattini, an Italian in Krakow, Poland, built three different models of a flying dragon—a most complex ornithopter. This machine seems to be one of the first to attract widespread attention among contemporary scientists, among them Huygens, the famous Dutch astronomer and physicist. Burattini is said to have made several actual attempts to fly, all ending in failure, but his flying machine was a favorite topic of conversation in court circles.[16]

1670 Francesco de Lana.[17] a Jesuit scholar of Ferrara, Italy, proposed an airship, usually referred to as the first rational design for a lighter-than-air craft and the first logical approach to the balloon problem. The ship was to be equipped with four large spheres, each twenty feet in diameter, made of thin copper. The spheres were to be evacuated of all

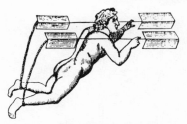

Besnier's flying apparatus,
from the Journal des Sçavans

A rendition of Burattini's
flying dragon

air, making them enough lighter than the surrounding atmosphere to lift the ship, which would be provided with oars and a sail for propulsion and control. Even if the principle was sound, the spheres would have collapsed under normal atmospheric pressure. While the project never materialized, historians regard it as a major milestone in aviation history, and the design is most spectacular.

1678 Besnier of Sable, France, a French locksmith, designed hinged wings stretched over frames and flapped by alternatively using both arms and legs. He did not claim to be able to take off from the ground, but thought that by starting from a height he could sustain himself in the air. The contemporary press was most sceptical, but his apparatus became known through an engraving in the technical *Journal des Sçavans* in 1678.[18]

1709 Lourenço de Gusmão, a Brazilian Jesuit priest who lived most of his life in Portugal, became widely known during the eighteenth century because of his fantastic flying vehicle, the "Passarola." His invention, like that of de Lana, was publicized by many engravings, some of them caricatures with the obvious intent of ridicule. No real original design seems to have survived, but the engravings, with their many variations, provide a fair idea of the intent of this interesting scholar and inventor. Gibbs-Smith especially devoted considerable attention to this design, and has ventured his own reconstruction of the original. It was to have had a boat-shaped fuselage, propelled and partly sustained by wings, with a bird's tail. A sail spanning the vehicle was intended to give constant lift to supplement the flapping of the wings. Like any full-scale, manpowered ornithopter, it was destined to remain earthbound. There is no record of further development of this interesting design.

Gusmao also gave attention to the idea of a hot air balloon; and indeed some historians hold that he was its actual inventor. He is said to have made one or more ascents on August 8, 1709, in the presence of the king of Portugal, who even awarded him a "patent" for the construction. His "balloon" was a small vessel in the form of a trough which

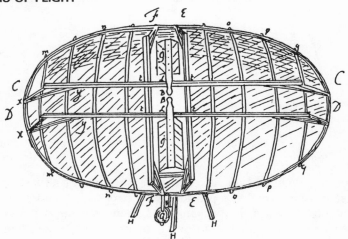

A rendition of Swedenborg's original sketch

was covered with a sail-like canvas. With various spirits, essences, and other ingredients, he put a fire beneath it, produced hot air, and enabled it to leave the ground. One version of the event says that it rose to a small height against a wall and then fell to earth and caught fire. Another version speaks of "a globe which rose gently to the height of the Salla—raised by a certain inflammable material which the inventor himself set fire to." Some historical records speak of the Inquisition becoming interested in Gusmao, although there is no evidence that he was ever prosecuted. He finally disappeared quite mysteriously. "Never before or since has there been such an interesting figure in the history of flying," according to Gibbs-Smith.[19]

1714-16 Emanuel Swedenborg, Swedish scientist, mathematician, philosopher, and religious thinker, designed a "vehicle to fly in the air." His original concepts included a fixed wing or hood with a propelling device to be operated by a man on top of the wing. The vehicle was never built or tried by Swedenborg; he was aware that with the sources of power then available, it could not fly. However, the later independent development of the fixed wing concept marks this design as a major advance in discovering the means to human flight.[20]

As was the case with da Vinci's designs, Swedenborg's became known only after others had come upon the same idea. This detracts not at all from its originality, and Swedenborg must further be credited with making his contribution accessible through careful and scholarly publication. To the present writer, it is wholly appropriate that the first sight that greets visitors to the Smithsonian's early Flight Room is a model of this craft.[21]

II

BACKGROUND

ENVIRONMENT, SOURCES OF INSPIRATION

Having set Emanuel Swedenborg's airplane design in its broadest context, we may now focus in more closely on the design itself, looking first at its genesis.

Swedenborg's range of interests was extraordinarily wide, and he applied his genius to such diverse fields as science, philosophy, and religion—all this in his "spare time," while serving capably as a civil servant and legislator.[1] His was a restless and inquiring mind, which led him through vast areas of the human intellect, and beyond. It is difficult indeed for ordinary people to comprehend his significant achievements in so many fields.

A few of the individuals inspired by Swedenborg's religious message have read into his early mechanical inventions proof of a divine calling, perhaps basing this on a remark he made in his later years.[2] There is, however, no claim of religious inspiration associated with the original design or with its published version. The design emerged at a time when the young Swedenborg was clearly absorbed in "worldly" interests, fascinated by scientific questions, putting his energies into the first steps toward establishing his personal reputation and securing an appropriate professional post.

We can also dismiss any speculations about his childhood acquaintance with angels—a favorite topic of his Pietist father—as a significant source of inspirations for his ideas about flight. As we shall demonstrate below, his speculations were prompted by the scientific environment he sought out during his first trip abroad between 1710 and 1715, especially in England.[3]

SWEDEN, A COUNTRY OF PROBLEMS

We must first have a quick look at the general political, cultural, and intellectual atmosphere in the Sweden of the early eighteenth century. At the outset, we should mention two pervasive characteristics. First of all, like other countries at the end of the seventeenth century and the beginning of the eighteenth, Sweden found itself in a period of transition in the fields of philosophy, religion, science, and higher education in general. The Age of Enlightenment, with its irresistible trend toward acceptance of natural science as the true basis of knowledge and its critical view of past dogmatic, "Aristotelian" theological thinking, was taking Europe by storm.[4]

The new winds had started to blow in Sweden in 1649 when Queen Christina[5] invited the French philosopher René Descartes (Renatus Cartesius) to visit. While his stay in Sweden was short,[6] his teaching and ideas took root and caused an upheaval in the intellectual world that lasted for a full century.

Emanuel Swedenborg was one of the bright young students who came to the University at Upsala at the beginning of the eighteenth century. Without losing his reverence for traditional values, he was intellectually strong enough to make a relatively smooth break with classical academic modes of thought. In this he had the strong moral support of older teachers and models, men of high learning and standing, some with international repute.[7]

Chief among these was Eric Benzelius (1675-1742), scholar and politician, son of one of Sweden's archbishops. He became one of the leading personalities in the nation's cultural life. He visited and maintained contact with such outstanding philosophers as Leibniz, Thomasius, and Malebranche toward the close of the seventeenth century (1697-1700). He also visited England and made a study of Oxford University. He became Chief Librarian of Upsala University in 1702, in which critical position he could further various sciences and acquire books and manuscripts for the university. He was Bishop of Gothenburg from 1726 to 1731, then Bishop of Linköping from 1731 to 1742. He did not take up the position of Archbishop, as he died before he could assume the position.

Benzelius married one of Swedenborg's sisters; and when Jesper Swedberg left Upsala for the bishopric of Skara in 1703, his son Emanuel, still a university student, was housed by his sister and brother-in-law, to whom he later referred as "a second father." His significance in Swedenborg's life can hardly be overestimated. In 1710, Benzelius took the initiative of founding the "Collegium Curiosorum," the country's first learned society, enlisting the most outstanding men of learning in Sweden. During his stay abroad, Swedenborg was a contributing member of this society, which, as the Royal Society of the Sciences in Upsala, is still an active institution in Sweden's intellectual life today. His letters to Benzelius, preserved in the Diocesan Library of Linköping, constitute one of the main sources of information about Swedenborg as a young man. They include a manuscript and drawing of the airplane.

The stature of the Collegium may be gathered from other contemporaries who were also members, including Carl von Linné (Linnaeus, 1707-78), the founder of systematic botany, Anders Celsius (1701-44), inventor of the Centigrade system, and Olof Rudbeck, Jr., botanist and linguist.

The other pervasive characteristic of Sweden during these years centered in the political, military, and economic situation. After the great victories of Charles XII[8] during the first years of the century, the country's fortunes turned. Charles' military campaigns in Russia ended in humiliating defeats which disillusioned many of Sweden's leaders concerning their monarch. He was a prisoner in Turkey for three years, returning in 1715 with his expansionist ambitions unbridled. His policies brought the country to the brink of political chaos and economic ruin before, in 1718, he died in battle from a sniper's bullet—perhaps from one of his own people.

It was then a turbulent world in the midst of which Swedenborg designed his airplane. When his invention was published in 1716 (he had had it in mind as early as 1711), who was there in the poverty-stricken country of Sweden to take interest in such a utopian notion? It was certainly an idea whose time had not come. Both the ruling classes and the impoverished masses would regard it as useless or even frivolous, unattainable, and perhaps even dangerous and suspect of sorcery.

CHILDHOOD AND STUDENT YEARS

With this overview of the cultural and political context, we may now turn to the personal context. The airplane design was an episode in a purposeful individual life, and it is appropriate to look at its sources and its effects. It is no easy matter to present a brief summary of a career which offers so many promising avenues for further exploration.

Emanuel Swedenborg (or more properly Swedberg at the beginning of our story) was the son of one of Sweden's most eminent clergymen, Jesper Swedberg (1653-1735), chaplain to King Charles XI, professor of theology at the University of Upsala, and bishop of the diocese of Skara. Jesper was a determined reformer, making significant contributions to a revised translation of the Bible, to a revised hymnal, and to the improvement of the educational system.

His son Emanuel did not follow in his footsteps. At the age of twelve, he began living with his brother-in-law Eric Benzelius, whose Cartesian interests seem to have fired the boy's imagination.[9] Emanuel had entered the university in 1699, at the age of eleven, and he was to "graduate" according to the rituals of the time in 1709, when he was twenty-one. His thesis—a kind of dissertation demonstrating academic maturity without giving any specific indication of professional position—demonstrated literary, philosophical, and linguistic skills.[10] It would soon become evident, however, that his main interests lay in such fields as mathematics, mechanics, and physics.

We should note that Swedenborg's university days coincided with the height of Sweden's fortunes. Charles' victories[11] touched off a fervor of jubilant nationalism. The young monarch was, however briefly, the dominant political and military personality in Europe. The Duke of Marlborough[12] came to him from England to enlist his support in the Spanish Succession war. The king was admired and adored by all, Swedenborg included.[13]

The turn in Sweden's fortunes came just before Swedenborg's graduation. Charles' defeat at Poltava and his subsequent imprisonment in Turkey marked the abrupt end of Sweden's eminence. Not surprisingly, popular sentiment in regard to the king swung swiftly to the opposite extreme, especially as Sweden's economic plight worsened.[14] An attack by neighboring Denmark was repelled,[15] but suddenly the plague spread through the North, and many who could afford to left the country. Swedenborg began making his own travel plans in 1709, though his departure was delayed.

SWEDENBORG ABROAD

The chaotic situation in Sweden, the uncertainty of the future, and above all the plague moved Emanuel to go abroad in 1710. He was determined to study and learn as much as possible in order to gain a worthwhile position on his return, and to make his own contribution to Sweden's hoped-for return to international prominence.

The trip itself was eventful, but that story lies outside the scope of this book.[16] He arrived in England armed with excellent letters of recommendation from his father and from Benzelius, which gave him entree at the highest levels of scholarship (See Appendix). He was to stay in England for more than two years, spending another year "returning" through the Netherlands and France, and reaching mainland Sweden early in 1715 after a five-month stay in Pomerania in Northern Germany, then a Swedish province.

To a brilliant and ambitious man of 22, England and its advanced intellectual community must have been a tremendous experience, enhanced especially by an unprecedented spirit of freedom of speech and of inquiry. His appetite for everything new was insatiable, especially in the fields of mathematics, physics, and mechanics, and he soon became equally fascinated by astronomy. While it is unlikely that he met Newton, he read him voraciously, and absorbed the revolutionary new ideas about gravity.[17] This gave new cogency to his studies of astronomy, geology, mechanics, optics, and chemistry. While his credentials gave him access to the eminent,[18] he made a practice of living with craftsmen. In this way, he acquired the skills of bookbinding, clockmaking, cabinetry, instrument making, engraving, lens grinding, and marble inlay. Although he never rose to the level of master craftsman, he apparently retained competence in some of these crafts in later years.[19]

Beyond this, he seems to have taken an avid interest in the study of past and contemporary literature in science and philosophy, devouring enormous numbers of books. At the request of the Collegium Curiosorum, he bought and sent home many major works which were unavailable in Sweden.[20] His travels in the Netherlands and France were marked by the same energy and the same access to eminent minds. We might mention, in view of his later interest in Sweden's international relations, that he was helped by Swedish diplomats wherever his travels took him.

THE TIME OF CONCEPTION

There is no firm indication of the precise date on which Swedenborg's interest in the subject of flying began. Based on available documents, it is reasonable to assume that it started during his stay in London, in 1711 or 1712, when he was about twenty-three years old. This was when he described himself as having "an immoderate desire for astronomy and mechanics," as he wrote to Benzelius.[21]

We have no evidence from Swedenborg himself about his sources for the design or about the way it evolved in his mind. This information may very well have been included among the books, notes, and diaries which he left in the care of a commercial agent in Hamburg as he was returning to Sweden in 1714. The agent undertook to have this material shipped to Sweden, but it was never heard of thereafter. This was a blow to Swedenborg personally, and a significant loss to historical research.[22] We may note also that some documents sent to his father in Skara were lost when a violent fire destroyed the bishop's home at Brunsbo in 1712.

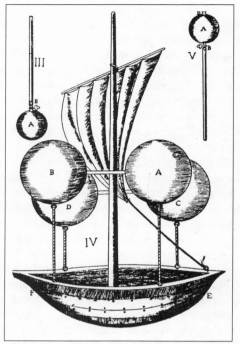

Francesco de Lana's flying boat

We are therefore left to speculate on the basis of probabilities, knowing his interests and the circles in which he moved. It is unlikely, first of all, that he had heard about flying in Sweden; so it may have come to him as a revelation when he suddenly found himself in the midst of inventors and experimenters in the mechanical fields, some of them preoccupied with the idea of flight. He had certainly heard stories about Francesco de Lana, who had designed the flying boat with copper vacuum spheres.[23]

There is no direct evidence of his acquaintance with the thought of the mystical figure of Lourenço de Gusmão[24] and his plan to trap hot air under a sail, but he can hardly have escaped the widespread attention given to this remarkable individual. He may well have seen engravings of this strange object, or have read about it in a London newspaper.[25] With Newton's speculation about gravity a prime topic of discussion in learned circles, it is entirely probable that ideas about flight also circulated widely.

We can surely say, with Carl von Klinckowström,[26] that the problem of flight had not been solved, and that Swedenborg had the kind of aggressive intellect that was drawn to "insoluble" problems. While he regularly availed himself of the researches of others, he was a disciplined and independent thinker;[27] and in the case of the airplane, there is no evidence of dependence on any previous design.

SWEDENBORG'S OWN REFERENCES

In the document we refer to as "the Manuscript," Swedenborg enumerates six points as "proofs" of the validity of his concept. This list is somewhat awkward in form, and may be used primarily as an index to the subjects which he had studied in order to reach his specific conclusions. He mentions the flight of eagles and "gleads,"[28] paper kites, primitive experiments with parachutes, how wind can blow open a door against the resistance of two men, and so on. It indicates that he gave the matter serious and persistent thought, trying to bring all relevant information to bear.

He does mention a few individuals by name. One of these is Anastasius Kircher (1602-1757), a German Jesuit renowned for his learning especially in the fields of mathematics and physics. His book *Magnes* was in Swedenborg's library, and in his *Ars Magna Lucis et Umbrae* (1646), he presented a detailed treatment of fixed-surface kites.[29]

Swedenborg also mentions Bernard le Bovier de Fontenelle (1657-1757), secretary of the French Academy of Science from 1697 to 1740. Later scholars have criticized him for writing in a light and popular style, his most respected work being *Entretiens sur la Pluralité des Mondes* (1686). In his published account of the airplane design, Swedenborg quoted Fontenelle's prediction: "The art of flying is only in its birth process; it will develop, and someday we will even travel to the moon. Do we think that we have discovered everything, or have reached the point where there is nothing more to add? Hah! Let us have the good grace to admit that there is something left for the centuries to come."

Swedenborg may have met Fontenelle during his stay in Paris in 1713-14; or he may have heard him lecture and, like others, been caught up by his skill as an orator. Certainly he would have felt akin to his imaginative and visionary mind.

Swedenborg does not mention Francesco de Lana[30] by name, but somewhat scornfully rejects the kind of design for which de Lana was known. We must presume that he had subjected the proposal to his usual scrutiny.

Swedenborg also mentions Christopher Polhem,[31] presumably for more than one reason. He was Sweden's foremost inventor, with an international reputation in the field of mechanics, and he had studied the resistance of air to falling objects. He was also in a pivotal position to advance Swedenborg's professional interests.[32] In his published account, Swedenborg was careful to note Polhem's objections to the proposed design.[33]

PUBLISHED SOURCES

Swedenborg had access in England to excellent libraries, and he took full advantage of them. His proposal shows that he was familiar with past failures and was on the track of discovering their causes. We have noted his explicit references to the work of Kircher, and should also call attention to his familiarity with the conclusions of G. A. Borelli, who in 1680 had demonstrated (although his calculations were incorrect in detail) that the weight-to-power ratio of the human body made it impossible for people to fly by their own exertions.[34] Swedenborg's agreement with this conclusion was probably the main reason he turned to a kite-like fixed wing surface—it would be easier to propel ourselves and to use a large lifting surface than to lift ourselves by main force.

Swedenborg's attention was also caught by the work of Robert Hooke (1635-1703), an English physicist, mathematician, and inventor, with such achievements as a coil spring, a quadrant, and a telescope to his credit. In a collection of Hooke's works published posthumously by R. Waller in 1705, there is mention of a design for a flying machine to be propelled by a kind of wind wheel, with a model built in 1658. The account gives no reason to believe that it would have been successful.

Bishop John Wilkins (1614-1672), who had seen Hooke's model, published his *Natural Magic* in 1648. It included material on the problem of flight, and may have provided the intitial stimulus for several of the inventions Swedenborg mentions in his 1714 letter to Benzelius.[35]

III

THE MACHINE:

"THE MANUSCRIPT" AND
"THE PUBLISHED ACCOUNT"

A LETTER OF SEPTEMBER 8, 1714

We may now turn to the design itself. It was unknown to scholars until Rudolf Tafel (an outstanding Swedenborgian researcher, Professor at Washington University in St. Louis, MO)[1] visited Linköping in 1867-68 and discovered the Benzelius collection of Swedenborg documents; and it was not until 1909 that English translations seem to have been available. The few treatments of this material have used it primarily to suggest the scope of Swedenborg's interests and abilities, generally noting his own awareness that the machine could not actually fly. It is only since World War II that there has been a disciplined interest in the history of aviation, the few exceptions being noted in the Bibliography.

The first documentary evidence we have for Swedenborg's design is a letter he wrote to Benzelius on September 8, 1714.[2] At this time, he was on his way back to Sweden. He needed to bring back evidence of his learning in order to further his career, so he stopped in Rostock[3] to put his achievements in presentable form. He mentions in the letter fourteen "mechanical inventions, either in hand or fully written out." Most of them were not too radical, and could have been realized within a short span of time; some indeed were already in existence elsewhere. But in Swedenborg's era, two of these inventions were regarded as impossible—the "submarine" (Invention No. 1) and the "flying vehicle" (Invention No. 12a).

Of these, only the second was in due time provided with a sketch, and it is likely that it would have received no attention whatever were it not for this later development. Even the description would not have

15

caught the imagination the way the sketch does, as the development of models attests.[4]

The complete list reads as follows:

1. A ship which, with its steersman, can go under water and do great damage to the enemy.
2. A syphon for quickly raising large quantities of water.
3. A method of lifting weights by means of water and the above mentioned syphon.
4. A method of constructing sluices where there is no fall of water, whereby ships can be raised to any required height within an hour or two.
5. A machine driven by fire for throwing out water.
6. A drawbridge to be operated from within the gates or walls.
7. New air pumps, also a pump worked by water and mercury.
8. An air gun, a thousand of which can be discharged by means of a single syphon.
9. A musical instrument, with the airs marked on paper, which can be played by one entirely unacquainted with music.
10. A method of engraving on any kind of surface, by means of fire.
11. A water horologue; showing the motion of the heavenly bodies by the flow of the water.
12. A carriage containing various mechanical contrivances set in motion when the carriage is drawn by the horses.
12a. A flying vehicle, or the possibility of being sustained in the air, and being conveyed through it.
13. A method of conjecturing the desires and affections of men by analysis.
14. New methods of constructing tensions and springs.

In the same letter, he says that he has nearly finished bringing his scattered notes into order, and promises, "In a short time I will forward to you the drawings at which I am now working daily."

He apparently forwarded drawings as he completed them, since in a letter dated Greifswald,[5] April 4, 1715 (shortly before his return to mainland Sweden), he writes, "By the last mail, I sent enclosed in my dear father's letter a drawing of an air pump to be worked by water.... In my last letter to my dear father I promised to send on every occasion ... one or two machines, being the result of my speculations, [asking him] to forward them to you.... I shall continue to do so for some time."

We do not know how many of these drawings were sent. However, among the Benzelius papers in the Diocesan Library of Linköping, we find in Swedenborg's handwriting "a description of a syphon machine for throwing water" and "A description of a machine for flying, with a drawing," which latter we refer to as "the manuscript." Except for a design for a new air pump published in the third number of *Daedalus*

Hyperboreus and "the published account" of the flying machine in the fourth number of the same, these are the only evidences for the fourteen inventions.

There are then two descriptions of the flying machine, the manuscript, which we can date only approximately to the fall of 1714,[6] and the published account from the October-December, 1716 issue of *Daedalus Hyperboreus*.

The Swedish in which the documents are written is, to the modern reader, distinctly archaic, and at some points presented particular difficulties to the Odhners as translators. I have chosen to reproduce their translations verbatim, with any improvements indicated in footnotes.[7]

"THE MANUSCRIPT"

Description of a Daedalian or Flying Machine

1. Make a square box, or car, t t t t, of lightest possible material, such as leather, cork, or, best of all, birch bark, with thin wooden splints, the whole, however, strong enough to hold a man without danger. The box should be 2 ells [yards][8] in length and 3 in in width, for the wings are to be moved sideways and the box, therefore, should be greater in width than in length; the depth should be 1 ell, with space within for a Daedalus or aviator.

2. A sail[9] should be stretched as wide as possible and be bent into a hollow; examine it carefully to make sure there is no rift or crack in the sail for the air to pass through, for the pressure or the atmosphere will be so strong that if there be any rift, the air will be forced through with a whistling noise. The sail should be 150 square ells, for since an eagle or glead occupies about 2 square feet of surface when it lies still in the air, with tail, body, and all, therefore, as our flying machine with all its gear will be about 300 times heavier than an eagle, it must also occupy a space 300 times greater than an eagle in relation to its weight, that is to say, 150 cubic [square?] ells. The sail may be made in the form of a square, oblong, circle, just as you please; best, perhaps, would be an oval form, provided with a support.

3. The sail is fastened and bent as may be seen from the drawing, [Fig. 2]. viz., there are 4 poles lengthwise and crosswise, CC, DD, EE, FF, all of which are bent at the ends, at yx, yx. Then make a thin wooden rim around it, D.C.m.n.F bent into an oval shape. To this rim are fastened the splints, m.m, n.n, o.o, which are also bent, and underneath all this the sail is securely fastened.

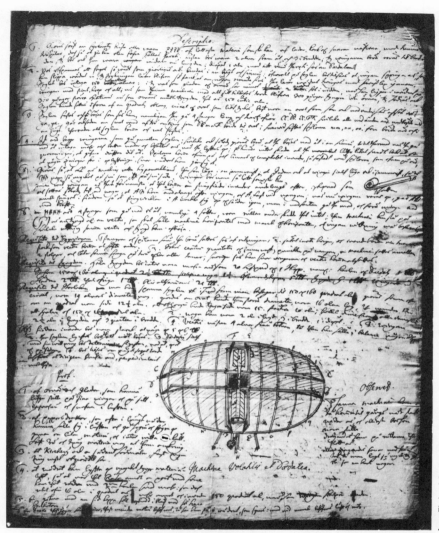

"The Manuscript"

Lars Ekelund, Linköping

4. The two wings, j.j, are placed between the sails, and they also should be bent into a hollow, in order to catch and keep the air better at the downward stroke than at the upward. It would do no harm if they were tilted obliquely backwards, like the arms of a wind mill. They may be made of splints covered with sail-cloth, so fastened to the splints above that it opens a little at the upward stroke, for the air to pass through.

5. L.L. is the true centre of the machine, where it lies in a balance. The wings must also be balanced, so that B B on the one side is equal in weight with J J on the other. The wings should be as light as possible, but the most important thing in connection with them is a spiral spring (*sling-fjaeder*),[10] placed lengthwise beneath each, shaped as in Fig. 1.

This spring is fastened to the car in such a manner that A B runs lengthwise along the wing, while A is fastened to the car itself. Now when the wing moves upwards, A. B moves against the spiral in the spring, so that the spiral circle in A is rolled up, but at the downward stroke it recoils and carries the wing downward with force.

6. H H H H are four poles proceeding downwards, *i. e.*, four legs; it would do no harm if they were provided with rollers,[11] for the machine to stand upon. G is a weight or beam [*vectis*], which is to keep the machine in a horizontal position, so that it may not tip over.

Requisites for the building of the machine: 1) The position of the poles and the splints may be seen from the drawing. 2) The seat is within the basket, and beneath there ought to be a bar under which the weight is fastened. 3) The center of gravity, or the balance, should be determined in the middle of the car,—the machine should be placed between two poles and hung on two axles, from which it may be seen where the wings and the weight are to be fastened.

Requisites as to the weight. The weight of the whole machine ought not to be more than 20 Lispund, or 1 Skeppund, viz., the man or Daedalus, 8 Lp.; the sails 150 or 160 square ells,—2½ Lp.; the car itself 1½ Lp.; the framework 5 Lp.; the weight G, 1 Lp.; the wings, with the springs, 2 Lp.; the rest 1 Lp. Altogether, 21 Lp. [The old Swedish "Lispund" = 18 lb. 12 oz. The Skeppund = 374 lb., or 171 kilogr.]

Requisites as to the size. The sails may be expanded in whatever form you please, to the size of 150 or 160 square ells. If the form is circular, then a diameter of 14 ells would be sufficient; if oval, then the larger diameter might be 16 ells, and the shorter 12; if square, then the side would be 12½ ells. If oblong, the longer sides ought to be 15, the diagonal 10 ells; all these dimensions would occupy a space of 150 or 160 square ells. 2) The car should be 2 ells long, 3 broad, and 1 deep. 3) The wings 2½ ells long, and ¾ broad. 4) The weight or beam should be almost 4 ells from the bottom, when it would be able to keep in balance several Skeppund.

Obs. The springs below the wings ought to be strong and should weigh about 5 or 6 Lp. 2) Where the sails are fastened [?] they ought to be bent inward. 3) The Daedalus himself should determine the flight by his swaying downwards, upwards, or to the side. 4) One must see if any [additional] sail be necessary to direct the course, downwards or perpendicularly.

Proofs

1. From eagles or gleads, which are able to lie still on their wings or on their expanse, or sway in the air.

2. From paper-kites, which often in calm weather are able to keep themselves in the air, and rise higher and higher up by only a light motion, and yet not tip over, although surrounded by wood and other heavy materials.

3. That Kirchberg[12] and others, tell about such things although nothing [additional] is seen expanded.

4. That the wind can lift up very heavy materials, so that when it blows against a gate with force, it can blow it open even though two men be pushing against it, when yet it is often 16 square ells in extent. How, then, would it act on a surface of 150 square ells, with the wings helping along?

5. A student with a side-cape fell unharmed down from Skara[13] church tower in a strong wind.

6. A kite, the higher it reaches, the less motion is needed to keep it in the air, as is apparent; while down by the ground it has to be lifted up by motion.

Observata

A machine such as this can be made to go when there is a strong wind; otherwise it will remain still.

It may be drawn forward on the rollers, where the ground is even. Or it may be pushed down from a roof, after it has been weighted with ballast, to the weight of a man.

DAEDALUS HYPERBOREUS

Swedenborg returned from his first journey abroad with several projects in mind. The one that came to fruition, in part through the support of Eric Benzelius, was the publication of a scientific journal. The thought had been in his mind as early as 1709, before he left for England; and the topic had been discussed by the Collegium Curiosorum. The project's acceptance was also furthered by Swedenborg's eagerness to disseminate information about Polhem's inventions,[14] a fact which was appreciated by both Polhem and the Collegium, and by his willingness to do the editorial work himself and to defray the costs of publication out of his own means.[15]

His travels had now convinced him that the time had come to raise the general level of knowledge and education in his own relatively backward and impoverished native land. He therefore reasoned that the journal should be published in Swedish and not in the "international language" of Latin. Polhem concurred in this decision, although the king advocated Latin.

Swedenborg had more in mind than benefiting Polhem and Sweden. He wanted an outlet for his own ideas, and saw the advantages of using Polhem's prestige to this end. Finally, with Charles XII home at last after his long exile, his royal favor was a major factor in professional advancement, and the young man was eager to prove his qualifications to the king. His creativity, diligence, and competence in originating, editing, and contributing to the first Swedish journal of its kind would promote his cause effectively if they came to the king's notice; and Swedenborg worked intensely to arrange that they should.

Six numbers of *Daedalus* were published betweeen 1715 and 1718; another three numbers were planned but never issued. Five hundred and fifty copies of the first number were printed, fewer probably for succeeding ones. The only likely international distribution would have been to Swedenborg's friends and to Sweden's foreign legations.

The description of the flying machine appeared in the fourth issue.[16] Polhem apparently saw the issue in draft form, since he communicated to Swedenborg not only his reservations about this proposal, but also his objections to changes in his own article on the resistance of air to falling bodies. Originally written in Swedish in 1711, it was translated into Latin in 1714; and in connection with this translation, Swedenborg, without Polhem's knowledge or consent, added some mathematical calculations of his own and apparently added some alternatives. Polhem's response was careful and polite, and Swedenborg is to be commended for publishing the eminent man's reservations about the airplane fully and fairly.

"THE PUBLISHED ACCOUNT"

[The following is the Odhner translation of the article in *Daedalus Hyperboreus*.]

Suggestions for a Machine to Fly in the Air

From the thought of Assessor Polheim [in a preceding article on "Rational Duplicates in Perpendicular Falls"], we may figure out the power and resistance of the air to all objects, expanded as well as compact or compressed. From this, as also from the flight of birds, it would be easy to come to the conclusion that a machine might be invented which could carry and transmit us through the air, and that we are not to be excluded from the element overhead, even though no other wings be given us than those of the understanding. Those who before now have given thought to such a work of Daedalus or Mercury, have set before them an impracticable principle, and have founded their notions on things contrary to our atmosphere, viz., great balls which, by being emptied of air, acquire a sufficient lightness to raise up a machine and its Icarus as well.[17]

But if we follow living nature, examining the proportions that the wing of a bird holds to its body, a similar mechanism might be invented, which should give us hope to be able to follow the bird in the air. First, let a car or boat or some object be made of light material such as cork or birch bark, with a room within for the operator. Second, in front as well as behind, or all around, set a widely stretched sail parallel to the machine, forming within a hollow, or bend, which could be reefed like the sails of a ship. Third, place wings on the sides, to be worked up and down by a spiral spring, these wings also to be hollow below in order to increase the force and velocity, take in the air and make the resistance as great as may be required. These, too, should be made of light material and of sufficient size; they should be in the shape of bird's-wings, or the arms of a windmill or some such shape, and be tilted obliquely upwards and be made so as to collapse on the upward stroke and expand on the downward. Fourth, place a balance or beam [*vectis*] below, hanging down perpendicularly to some distance and with a small weight attached to the end, pendent exactly in line with the center of gravity,— the longer this beam is, the lighter it must be, for it must have the same proportion as the well known [Roman] *vectis* or steelyard. This would serve to restore the balance of the machine whenever it should lean over to any of the four sides. Fifth, the wings would perhaps have greater force, so as to increase the resistance and make the flight easier, if a hood or shield were placed above them, as is the case with certain insects. Sixth, when now the sails are expanded so as to occupy a great surface and much air, with a balance keeping them horizontal, only a small force would be needed to move the machine back and forth in a circle, and up and down. And after it has gained [sufficient] momentum to move slowly upwards, a light movement and an even bearing would keep it balanced in the air and would determine its direction at will.

It seems easier, however, to talk of such a machine than to put it into actuality and get it up into the air, for it requires greater force and less weight than exists in the human body. However, there are three or four requisites that would be of chief assistance: 1) a strong wind, which has a considerable effect on similar objects, for in calm weather it would be better to keep quietly and humbly by the ground; 2) the machine should be pushed off from a considerable elevation, for the primary difficulty will be to force oneself up from the level; much would be gained toward this end if the machine were lifted up some distance into the air by means of ropes; this would do as much as a strong puff of wind; 3) the size, and width of the sails, (as well as the force of the wings), ought to be in proportion to the weight, and must increase with the weight in the ratio of 3 to 2, as Assessor Polhem shows in the article below; 4) in order to acquire a downward force in the wings sufficient for moderately calm weather, the science of mechanics might perhaps suggest a means, viz., a strong and stiff spiral spring, which, when set free, would have the power of three or four persons; and it could be bent upward, though somewhat more slowly, by a light and quick mechanism.

If these advantages and requisites are observed, perhaps in time to come one might know how better to utilize our sketch and cause some addition to be made, so as to accomplish that which we can only suggest. Yet there are sufficient proofs and examples from Nature that such flight can take place without danger; such as birds, eagles, and gleads, which, as it were, swim in the air and with all their weight rest on their wings without moving the least feather for several minutes. In the case of kites, made of paper and wood, we see a similar property, in that they keep themselves up in the air without sinking down in the least. It is also well known that a man in Strängnäs[18] accidentally fell down from a tower in a strong wind, but the cape which he wore so far saved him, that he came to the ground unharmed. And there are other cases which may be considered; although when the first trials are to be made, you may have to pay for the experience and must not mind an arm or a leg.

The ingenious Fontenelle writes humorously about such a machine, saying: "The art of flying is as yet hardly born. It will be perfected and some day people will fly up to the moon. Do we pretend to have discovered everything, or to have brought our knowledge to a point where nothing can be added to it? Oh, for mercy's sake, let us agree that there is still something to do for the ages to come?" (*Entretiens sur la Pluralité des Mondes*, pp. 51, 52)[19]

The learned Assessor Polheim[20] pronounces a more doubtful opinion, as follows: "As to artificial flying it probably has the same difficulties as the artificial production of gold or perpetual motion, etc., although at first sight it seems no less practicable than desirable; but when we examine the matter more closely we meet something which Nature seems to deny us; as in the present case that all machines do not retain equal proportions in great dimensions as in small, even though all the parts be made the same and in the same proportion; for instance, although a cane or stick may be capable of carrying not only itself but also some weight in addition, yet this does not apply to all dimensions, even if the length and thickness keep the same proportions, for when the weight increases in triplicate ratio, the force increases only in duplicate; it is the same as to surfaces. Whence it happens that great bodies cannot support their own weight. And since Nature itself demands of birds not only a very light and strong material for feathers, but also wholly different tendons and bones in the body itself, which tend to strength and lightness, and which do not exist in other bodies; therefore, on account of the lack of necessary materials, it is much more difficult to accomplish that effect in the air which is needed for the realization of this thing, in case a human body is to accompany the machine. But if it were possible for one person to move and direct all that pertains to a machine sufficiently large to be able to carry him, then the point would be gained. However, it would be well to take advantage of the wind, if the same were constant and invariable."

But enough this time, about our Daedalus.[21]

POSTSCRIPT

If the specific proposal for a flying machine was unproductive, the general effort of which it was a part accomplished its ends. *Daedalus* did come to the attention of the king, and it did provide an arena in which the relationship between Swedenborg and Polhem developed.[22]

On his return from Turkey, Charles found a country longing for peace and many subjects bitterly resenting the monarch who had led them into disastrous wars. Feelings ran so high that he never did return to Stockholm, the capital city, but made his headquarters in Scania in the South, first in Ystad on the Baltic coast, and later in the cathedral city of Lund. His reputation was not enhanced by his immediate efforts to raise money[23] and a new army to fight against Norway. But he was, after all, the absolute ruler, with dictatorial powers.

As we have noted, royal favor was essential to advancement. Swedenborg's talent was in his favor, but his youth and lack of entree to the establishment were formidable obstacles. The establishment in fact regarded him (not without some reason) as "pushy."[24] As he marshalled the support of Polhem, Benzelius, and eventually the king, the aristocracy in particular reacted with apparently angry opposition.

Bishop Swedberg was also active on his sons' behalf. On April 25, 1716, he wrote to Charles asking for the ennoblement of his three sons Emanuel, Eleazer, and Jesper, as was traditional for the sons of bishops. This would remove a major obstacle, since only the nobility could hold any but the lowest positions in government. This request was granted only in 1719 by Queen Ulrica Eleonora, after Charles' death.

In the meanwhile, however, Polhem stood high in Charles' favor, and in 1716, Swedenborg was offered the opportunity to travel with him to Lund to meet with the king. Immensely excited by the prospect, he immediately had the first four issues of *Daedalus* bound in a single presentation volume, for which he composed an effusive dedication to Charles.[25]

The dedication is clearly his effort to win the king's favor for himself. He tells of his own "small investigations and observations," which he "lays down in deepest submission at the Majesty's feet because of the gracious solicitude the King is pleased to show, in particular to mathematical studies . . . ."[26]

He continues with reference to Polhem's inventions:

Some of them have been described in this little work [*Daedalus*], and I have added the observations of other learned men, your Royal Majesty's subjects, together with my own investigations, which by most earnest reflection, I have sought to mature, both at home and also during a five years' journey in foreign lands where studia mathematica are most cultivated and are in the highest esteem.

This is merely a beginning, most gracious King; much more still remains hidden away which, presumably, will contribute great advantages to your Majesty's kingdom, especially in the development of manufacture, navigation, artillery, and the art of shooting.

If this work wins your Majesty's grace, it will certainly rouse up many other men, in submissiveness to lay bare their thoughts, and to offer them for your Royal Majesty's gracious pleasure. I remain to the hour of my death Your Majesty's, My ever gracious Kings's, most humble and faithful subject, Emanuel Swedberg.

The king received Polhem and Swedenborg several times in Lund late in 1716, and their meetings continued during the next two years. On one occasion, the presentation copy of *Daedalus* was lying on the king's desk, and it was clear that he was familiar with its contents. While Charles was primarily interested in technical development for military purposes—dry docks, the Göta Canal, and the transportation of warships overland—he apparently had a lively interest in science for its own sake, and particularly in mathematics. He had his own suggestions for a new arithmetical system, and evidently found a kindred spirit in the imaginative Swedenborg. There are some detailed and interesting reports on these conversations—with no mention whatever of the flying machine.[27]

These efforts, and in particular Polhem's recommendation,[28] secured for Swedenborg a royal appointment as "Extraordinary (i.e., unsalaried) Assessor" of the College of Mines, the governmental board charged with the oversight of Sweden's vital mining industry. The then members of the College resented this appointment of a young man who had not climbed the established ladder of preferment, and refused to seat him. Fortunately, the appointment also provided official standing as an assistant to Polhem, and Swedenborg was shortly busy overseeing the dock works at Karlskrona. It would be some years before he began to participate in the College of Mines, eventually to become one of its most respected members.

In any case, the story of the airplane seems to have ended. Perhaps Polhem's objections discouraged him, perhaps his interest was simply displaced by his new professional involvement, or perhaps he realized that Sweden's meager resources needed to be devoted to efforts that would bring an immediate return.

This ending, however, should not obscure the intrinsic merits of the design itself. Tord Ångström, a pioneer in Swedish aeronautical engineering, offers the following summary.[29] "Christopher Polheim," he says, "who was a statesman and realist, took a rather strong and critical view on Swedenborg's suggestion for a flying machine. But even if he, from a narrow point of view, had sound objections, it is the imaginative seer, befitting a clear-sighted genius (Swedenborg) to whom posterity will give laurels and honour. A more positive attitude from Polhem and a more open eye for the many merits of Swedenborg's suggestion could have been a starting point for a successful continuation of Swedish research and constructive results in this field. Time after time such kind of scepticism, represented by Polhem, hampered the development of promising ideas, while, on the other side more erratic and fancy ideas often have been met with undeserved encouragement."

IV
AFTER SWEDENBORG

THE REMAINDER OF SWEDENBORG'S CAREER

Swedenborg's ennoblement[1] was followed by his active participation on the College of Mines, where he eventually became a salaried assessor; and it entitled him, as Jesper's eldest surviving son, to a seat in the House of Nobles. He was a diligent member of the College, and in particular made extensive trips abroad to bring back to Sweden the latest technology in mining and smelting. He was equally active in the legislature, showing a serious concern for the country's ravaged economy. He argued persuasively for a policy of fiscal conservatism and industrial development at times when "quick fixes" had a strong appeal. He also wrote extensive works in philosophy and anatomy which were very favorably reviewed abroad.

During a visit to the Netherlands in 1743 he entered a profound religious crisis which lasted for the better part of two years. His life was radically changed by this event, and over his last twenty-seven years, his writing was devoted almost exclusively to the theology for which he is most widely known.[2]

His posthumous reputation has been divided, leaving him a particularly controversial figure. Opinions range from regarding him as a truly inspired messenger of God to seeing him as completely mad.[3] His theology did find adherents, and small but devoted Swedenborgian churches grew up in many parts of the world. His philosophical and religious thought became a source of inspiration to many intellectuals, especially in the field of literature,[4] and he remains one of the most widely known Swedes of all times.

NEXT STEPS TOWARD FLIGHT

It was not until the middle of the eighteenth century that a noticeable interest in the possibility of flight arose. The first successful balloon ascent in 1783 tended to focus energies on lighter-than-air craft,[5] though most interested individuals still kept a foot in the heavier-than-air camp. For the most part, they disregarded Borelli's conclusions and continued to experiment with variations on the flapping wing, the ornithopter method.

Swedenborg, by taking Borelli seriously and by reflecting on the *gliding* of birds and the force of air against fixed surfaces, moved directly to the concept of the fixed wing. Gibbs-Smith leaves open the possibility that Melchior Bauer (1733-1???) may have been aware of Swedenborg's design when he produced a fixed-wing monoplane in 1764, but there is no documentary evidence for this.[6]

We may note in this connection that inventors tended to be secretive about both their discoveries and their sources, in order to gain full and exclusive rights to the benefits of any advances. The tendency increased with the passage of time, and the Wright brothers were involved in serious and complex patent controversies in the years after their first flight. The tendency to secrecy obviously complicates the task of the historian.

THE ROAD TO SUCCESS

Most historians label Sir George Cayley (1773-1857) as "the father of the airplane," and with good reason. He compared the effectiveness of different approaches, experimenting with a wing with a rotating arm and a fixed-wing monoplane, and tried to prove what Swedenborg seems to have assumed, that a curved wing surface generates more lift than a flat one and that the critical factor is the air pressure underneath. He published his findings in *Nicholson's Journal of Natural Philosophy* in 1809-10, setting a standard for the dissemination of information about the problems of flight.[7] He also made a determined effort to learn about past experiments[8] and mentions Wilkins, who was known to Swedenborg, but not Swedenborg himself. We may mention that he was a deeply religious man who in many ways shared Swedenborg's neo-Platonic views about the correspondence between material and spiritual things.[9]

William Samuel Henson (1805-1888) and John Stringfellow (1799-1883) carried on Cayley's tradition. They were the first to design a machine which used steam power. Henson, an ardent student of past experiments, regarded himself as a developer more than as an inventor, but his contagious enthusiasm and energy combined effectively with the engineering skills of Stringfellow. The two had a strong influence on further developments in Europe, especially through their rather fantastic 1843 design for an "Aerial Steam Carriage."

The fixed-wing concept reached one bizarre extreme in the design of Francis H. Wenham (1824-1908), who was one of the driving forces in the founding of the Royal Aeronautical Society in London in 1866.[10] During the years 1858-59, he carried out extensive experiments with a glider having no fewer than five fixed wingdecks on top of each other! In 1866, he delivered an influential lecture to the Society, presenting ideas about the curved wing profile that strongly resemble Swedenborg's.[11] It is also suggested that his basic concept was of a fixed airfoil propelled by flapping wings or by some kind of oars.

OTTO LILIENTHAL

As we approach the end of the nineteenth century and the achievement of heavier-than-air flight, we come to one of the greatest of aviation's pioneers, the German Otto Lilienthal (1848-1896).[12] He became the dominant figure in the field, and must be regarded as the immediate predecessor of the Wright brothers.

He and his brother Gustav studied, read, and experimented constantly. Like many others, they made careful observations of the flight of birds; Lilienthal's 1889 work *Bird Flight as Basis of Aviation*, though only three hundred copies were printed, became a classic. While he did not wholly abandon the ornithopter idea (one design was for a glider with flapping wingtips), his most intense energies went into the design and building of fixed-wing gliders, which he tested from a hill he constructed near his home in Berlin-Lichterfelde. He also made some experiments with models powered by a carbon-dioxide engine.

Lilienthal was fatally injured at the Gollen-Berge airfield near Berlin on August 9, 1896. His back was broken when a wind gust turned over the plane he was testing, and he died the following day. He is rightly remembered as the man who brought the infant technology to the very brink of success. His primary legacy, of central importance, was the aerodynamic superiority of the curved wing surface. [13]

THE AMERICANS

The immediate beneficiaries of Lilienthal's legacy were Americans, the principal figures being Octave Chanute (1832-1910) and his assistant

Augustus M. Herring, Samuel P. Langley (1867-1912), and particularly the Wright brothers, Orville (1871-1948) and Wilbur (1867-1912).[14]

Chanute[15] was one of the Wright brothers' principal advisors, an out-standing constructor of gliders and a student of his craft, seen by many as the first historian of aviation. His influential 1894 essay "Progress in Flying Machines" was a systematic, scientific, and authoritative presentation of the achievements in aeronautics to that date. It identifies by name 85 individuals: Swedenborg is not among them, Tafel's discovery in Linköping was not yet widely known, though in Glenview (a suburb of Chanute's Chicago) there were individuals who were very much aware of this design.[16] While he could be critical of Lilienthal's glider principles, he was the principal channel by which the German's progress was made available to Americans.

Langley was a specialist in mathematics and physics who became chief of the Smithsonian Museum in 1887. He is particularly noteworthy for his experiments with engine-driven propellors. His efforts very nearly met with success, but his rather monstrous plane, "The Aero-drome," toppled sideways when it was catapulted from a houseboat on the Potomac River on October 7, 1903, and slid into the water "like a handful of mortar."[17]

It was then on December 17, 1903, that success was finally achieved, with the Wright brothers' machine making the first manned and con-trolled heavier-than-air flight in history.

Their success can be seen to rest on two central factors. One was *the availability of an engine with a favorable ratio of power to weight*, and the other was *the use of fixed wings with a curved profile*. Swedenborg was fully aware of the need for the first, but had no solutions for the problems it posed. He was the first individual we know of to recognize the merits of the second. His design, then, represents a conceptual breakthrough of major importance, one of those strokes of genius which could have led to great things if others had become aware of it and had recognized its promise.

V

EXPERTS' EVALUATIONS

EARLY REACTIONS TO SWEDENBORG'S PROPOSAL

It seems that Swedenborg received no encouragement to pursue his interest in flight. Sweden's primary authority in the general field of mechanics was Polhem, and as we have seen,[1] he expressed his reservations quite directly. While King Charles had a lively interest in scientific and mathematical discoveries, he also had other things on his mind, and there is no suggestion that his discussions with Swedenborg touched on the subject of flight. No references to the published proposal have been found in contemporary Swedish or European sources, so we must view the immediate response as negative—either silence, or discouragement.

REDISCOVERY

The first affirmative attention to the design grew out of an interest not in the history of flight, but in theology. It was Rudolph Tafel's search for documents concerning Swedenborg, prompted by his adherence to Swedenborg's theological concepts, that unearthed the manuscript in Linköping, and it was Swedenborgians, especially in the United States, who first tried to test and evaluate it. The founding of the Swedenborg Scientific Association in Philadelphia, Pennsylvania, in 1898 provided the first forum for the sharing of such interests; and the first major step outside Swedenborgian circles was the whole-hearted engagement in 1902 of the Swedish Royal Academy of Science in the study of Swedenborg's life and works.

The Odhners' English translation of the manuscript was published in

Bryn Athyn, Pennsylvania, in 1909, and their translation of the published account was published in Philadephia in 1910. Most significantly, this latter was reprinted in the July, 1910, issue of *The Journal of the Royal Aeronautical Society* in London, with a foreword by the Editor, T. O'B. Hubbard. It was one thing for the Odhners to state in their introduction that "as far as it is known, Swedenborg was the first discoverer of the aeroplane." This might be dismissed as a partisan comment from individuals already convinced of their author's genius. It is quite another thing for Hubbard to describe the design as " . . . the first rational proposal for a flying machine of the aeroplane type . . . ." This must be regarded as the first fully professional evaluation of Swedenborg's design.

In 1916, Count Carl von Klinckowström published a summary and evaluation of Swedenborg's proposal.[2] He described it as a kind of glider without an engine, equipped with a pendulum stabilizer, to be controlled by the movement of the pilot's body weight. The wings might provide some lift and some forward motion; they could also be used as a steering device at higher altitudes. The curvature of the sail would enable it to serve as a parachute. He agreed with Polhem that the essential ingredient would be the wind,[3] and doubted that Swedenborg had been able to treat his invention mathematically.[4]

TORD ÅNGSTRÖM

One of the most enthusiastic students of the Swedenborg design in subsequent years has been the Swedish aeronautical expert Tord Ångström.[5] In the twenties and thirties, he wrote several articles on the subject. According to him, the most significant features of the design were:
- A fixed surface (the sail) for support
- A device for stability
- Provision for steering
- "Wings" for propulsion
- A cockpit for the pilot
- Landing gear

THE SMITHSONIAN NATIONAL AIR AND SPACE MUSEUM

The most thorough evaluation thus far was made on the occasion of the presentation of the model to the Smithsonian Museum in 1961.[6] The study was authorized by the Curator of the Air and Space Museum, Paul

Garber, and is of particular interest for the care with which it relates Swedenborg's suggestions to later developments.[7] It is worth citing in full.

> The Swedenborg machine of 1716 (?) is remarkable in that he so early conceived of a glider (a machine having a gliding supporting surface adapted to be launched from a height, drawn by a rope, flown on the wind, or raised by manpowered flapping wings) embodying several features commonly thought to have been used first by inventors of the Nineteenth Century.
>
> The gliding wing of *Sir George Cayley* reportedly used by him in 1809 had a section concaved longitudinally (in the direction of flight), but not also transversely as is the supporting surface of Swedenborg.
>
> The gliders of the German pioneer *Otto Lilienthal* and of the Americans *Octave Chanute* and *Augustus Herring* also had gliding surfaces cambered in the direction of flight. They experimented in the 80's and the 90's.
>
> Notable too is Swedenborg's early use of a pendular mass to assist in balancing the machine. Each of the Nineteenth Century gliders mentioned did this by suspending their own bodies pendularly from the supporting surfaces.
>
> Thirdly may be mentioned Swedenborg's concept of landing and launching struts which he says may be provided with wheels. Possibly the earliest such concept of the Nineteenth Century is that of the Englishman *Henson* about 1849.
>
> A fourth point of note is Swedenborg's use of two independently reciprocable lifting and propelling laterally extending wings. Because they are independently manually operable they might be considered as embryonic ailerons[8] in association with cambered main supporting surfaces. The Frenchman *Goupil* did not propose laterally extending ailerons until 1884.[9]
>
> Swedenborg's machine is longer than it is wide. Some other early inventors like *Mattulath* (who filed a patent application in the late 90's) also believed in the long narrow wing.[10] In the Twentieth Century the *Wright brothers* first sucessfully flew. Following them until quite recently airplanes have had wings of great breadth and small depth, just the opposite of those of Swedenborg and Mattulath. However, with the advent of supersonic airplanes and missiles long narrow wings are appearing. While these vehicles use wings almost if not quite flat in camber, and triangular in plane rather than elliptical or retangular, the early dreams of Swedenborg, Mattulath and others like them that a supporting surface should have its greater dimension in the direction of flight, one perhaps may regard as subject to prophetic interpretation.

EXPERT VIEWS OF 1986

In the quarter century since this last opinion, aircraft design has taken major strides. Is there more to be said about Swedenborg's design as a result? The author has put this question to two Swedish friends who are eminent professionals in the fields of aircraft construction and airworthiness.

Hans Eric Löfkvist, retired Executive Vice President of Svenska Aeroplanaktiebolaget, SAAB, Linköping, involved in engineering and development since 1939, responds as follows, after study of the documents themselves and of subsequent comments:

> When making a detailed study of Swedenborg's Suggestion for "A Machine to fly in the air" with an awareness of the level of knowledge at the time, one has to admire the methods he uses in order to attack the problems through sound technological application, both in its totality (= system solution) and in details. The most important innovation lies in his discarding, for technical reasons, the all too obvious thought of imitating birds and their self-propulsion by flapping wings, which both Leonardo da Vinci 200 years earlier and many later inventors, including Otto Lilienthal, had used as a starting point.

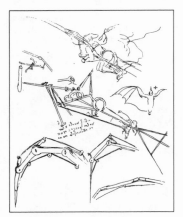

da Vinci's sketches for the ornithopter mechanism

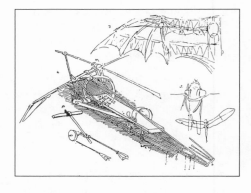

da Vinci's sketches for a prone ornithopter

> Since he can think of no other energy source than man himself and realizes that that source is insufficient, he sees the possibility of exploiting the forces in the air with a fixed carrying surface, thereby avoiding difficult problems from a construction point of view. The dynamics of a bird's wing and that of a mill have been transferred into a separate arrangement for maneuvering and propulsion which produces a *minor addition* to the primary lifting power of the air. With the assistance of a mechanical

refinement in the way of a spring he thinks that man power will be sufficient for this limited need.

Once the basic idea had been formulated some rather trivial construction questions remain. The structure of the fixed carrying surface, "the sail," is well thought out. A low weight is aimed at.

Swedenborg's knowledge about aerodynamics is totally lacking. This could have led to catastrophic consequences in two ways if he had tried to fly his machine—just as it happened to many other experimenters later on.

1. Like his contemporaries he did not understand that the "aspect ratio" of a wing (the ratio of span to mean chord) must be large to give the wing good performance in a gliding flight.

2. Likewise he had no knowledge of the distribution of the airload or air pressure over a wing nor, consequently, that the (imaginary) "center of pressure" varies with the angle of attack of the wing. The interplay between the center of pressure and the "center of gravity" of an aircraft is of vital importance for its longitudinal stability in flight, i.e. the ability to respond automatically to and counteract a disturbance of the flight position taken.

Today we know of several methods to achieve control and stability around the three axes of an aircraft, but the way was long and hard for the pioneers, with many crashes involved, sometimes fatal.

Birger Holmer, Vice President for Aircraft Research and Development, Scandinavian Airlines System from 1960 to 1981, now retired, has studied the design from a somewhat different point of view. On the basis of a careful examination of the Swedish text[11], he notes a discrepancy between the description and the drawing in regard to the orientation of the cockpit and the direction of travel as indicated by the placement of the "wings."[12]

He comments: "It is logical in order to get the highest possible strength from the pilot that the arm of leverage is as long as possible inside the pilot's place; he also ought to be able to stand between the handles if the balance so requires—according to Swedenborg's own reasoning on this point."

Holmer also finds some lack of clarity in regard to the position of the "spring"—its placement in the design does not correspond to the description in the text. This discrepancy had also been noted earlier by others, with various solutions proposed. The difficulty, according to Holmer, is that the spring in the small sketch is pictured upside down. This is apparently because when Swedenborg later made the large drawing, he realized that the design would look cleaner from a construction point of view if it followed the text, and therefore abandoned the orientation adopted for the smaller sketch.

The most serious error in regard to the translations has to do with the equivalence of the *aln*. The Swedish *aln* is approximately 0.6 meters—slightly more than half the English "ell," which is 1.14 meters. To Holmer's knowledge, only Clive Hart was aware of this, and all other commentators have therefore exaggerated the size of the machine.

Once this is realized, according to Holmer, Swedenborg seems to be correct in his ideas about the machine's structure, including its weight. The strength of the materials proposed is also adequate, and it should be able to support a man.

As to the "wings," Holmer agrees with Löfkvist, against Ångström, that they were intended to be moved up and down, not forward and back like paddles. With the additional force given them by the springs, and "tilted obliquely backwards," they will give both an upward lift and a forward push. One needs only rotary motion to arrive at the basic concept of a modern propellor—an ideal which does not seem to have occurred to Swedenborg.

Holmer also agrees with Löfkvist that the balance weight suspended below the pilot's seat will provide only a static equilibrium. Dynamic balance during flight has apparently not been thought of at all,[13] or at best is only vaguely dealt with by indicating that the pilot may control the craft's attitude by shifting his weight.

The machine should then, in Holmer's judgment, be regarded as a glider, with the "wings" giving some slight additional lift, some forward motion, and perhaps some lateral stability.

In summary, Holmer agrees with the common assessment that Swedenborg was indeed the first to recognize the necessity of a fixed carrying surface, which would need to be supplemented by a separate means of propulsion stronger than human muscles alone, and would also need to be provided with a carefully designed means of stabilization. For all its faults, the design represents historically a giant step in the right direction.

VI
DOCUMENTATION

THE SEARCH FOR THE SWEDENBORG DOCUMENTS

For all practical purposes, Swedenborg's airplane design was inaccessible to aviation historians until early in the twentieth century. True, there were copies of the account published in *Daedalus* in 1716, but this was in old Swedish in a virtually unknown journal. The manuscript, also in old Swedish, and handwritten as well, lay hidden in the archives of the Diocesan Library at Linköping.[1] Only when Swedenborgians undertook an intensive search for information about their theologian did this material begin to become available to a wider readership.

This search began shortly after Swedenborg's death in 1772. Perhaps primarily because Swedenborg's theology is intellectually challenging, the membership of Swedenborgian churches (especially in England and the United States) has tended to have a high regard for intellect and education, and has included many able individuals who have given generously of their professional knowledge and talents to the search for Swedenborg documents. Their efforts have in turn prompted non-Swedenborgians to continue the task.

During the first three decades after 1772, no fewer than four catalogues and descriptions of Swedenborg's works were issued—in 1772, 1782, 1787, and 1790. The last of these was a list compiled by the Swedish Academy of Science, and was revised in 1841.[2]

The primary researchers in the middle of the nineteenth century were the Englishman James John Garth Wilkinson[3] and the German-American brothers Immanuel[4] and Rudolf Tafel. Rudolf, the younger of the two, came to Sweden late in the 1860's and visited libraries and archives containing relevant documents, coming to Linköping in 1868. Among his finds were, of course, the manuscript and the letter of September 8, 1714,

37

Emanuel Swedenborg, 1688–1772

with the list of inventions. In 1869 he published a preliminary report, *Results of an Investigation into the Manuscripts of Swedenborg*, and within the next decade produced his definitive edition of the material.[5]

THE SWEDISH ROYAL ACADEMY OF SCIENCE

Understandably, the material thus discovered spread more quickly among Swedenborgians than among the general public. It is interesting to note that these documents came to light at the height of the activities of such aeronautical pioneers as Otto Lilienthal, just after the founding of the Royal Aeronautical Society in London, and not long before Octave Chanute began his research.

The first non-Swedenborgian body to take note of the discoveries, however, was the prestigious Swedish Royal Academy of Science.[6] In 1902, realizing that some of Swedenborg's discoveries were of major significance in the history of science, the Academy appointed a special committee to examine and evaluate his work. It was a highly qualified committee, well suited to make a new and objective study.[7] It was ultimately to publish three volumes,[8] which are still the primary resources for information about Swedenborg as a scientist. The first volume, published in 1907, contains the *Daedalus* proposal for the flying machine.

The secretary of this committee was a young American named Alfred H. Stroh, who lived and worked in Sweden for twenty years.[9] The indefatigable Stroh was a prime mover in bringing Swedenborg to public notice. He was instrumental in having Swedenborg's remains brought from England (where he had died) in 1908, and interred in a marble sarcophagus in the cathedral at Upsala, with the Swedish parliament bearing the costs. He edited the three volumes just mentioned; and perhaps most significantly, he was one of the main organizers of the international Swedenborg Congress held in London in 1910 under the patronage of King Gustav V of Sweden. Some 400 intellectuals from fourteen countries participated, including an impressive delegation from the Royal Swedish Academy of Science.

THE SWEDENBORG CONGRESS OF 1910

When this Congress took place, the Odhners had just finished their translations of the manuscript and the published account. The book on the machine was on display; and more than that, so was a model of the craft built by the Rev. J. A. Rendell, an Englishman who was Chairman of the Congress' Committee on Science.

The beginning of Polhem's article in the October-December, 1716 issue of Daedalus

The beginning of "The Published Account"

The model was unveiled at a garden party, and could scarcely have come at a more propitious time. New developments in flying were coming thick and fast, and were making the headlines.[10] The popular imagination was caught up in the excitement. The first flights in Europe came in 1906-8; in 1909 Louis Bleriot made the first flight across the English Channel; in 1910 air passenger service started, by Zeppelin, and the first woman pilot was licensed.

In this atmosphere, the unveiling of Rendell's model could hardly fail to stir excitement and pride. The design may look strange to us, accustomed as we are to sleek metal craft; but in comparison to the boxy airplanes of that era, cluttered with struts and wires, it must have looked quite different.

It did indeed attract attention outside Swedenborgian circles. We have noted its very favorable mention in the *Journal of the Royal Aeronautical Society*, but the news did not stop there. The event was also noted in the Swedish newspaper *Dagens Nyheter* on July 3, 1910, with a description of Swedenborg's design. This seems to be the first report of the machine to reach the general public in Sweden.

VII

"THE MACHINE"
IN AVIATION LITERATURE

THE BEGINNINGS OF PUBLIC NOTICE

In this chapter, we will focus on the spread of knowledge of Sweden-borg's design through the years. We may observe at the outset that two problems face the aviation historian in this regard. The first is that library indices have no separate listing for this topic. The information is buried in masses of material, either about Swedenborg or about aviation history, and few libraries have extensive collections in both fields. The second difficulty is that for most researchers, the Odhners' translations[1] are the closest available approximations to primary sources.

The result is that, while many individuals have mentioned or dealt with the machine in various contexts, the literature has a distinctly piecemeal cast. The pieces have not been put together; the puzzle has not been completed. The present book is an attempt to fill this gap.

Clearly, public notice starts with the Odhners' publication and espe-cially with the International Congress in 1910. The immediate favorable attention by the Royal Aeronautical Society in England is indeed impres-sive. However, public interest tended to be held by the succession of new developments. There was no lack of heroes, of ingenious and dar-ing pioneers. Further, there was extensive discussion of the possible use of aviation in warfare, a subject on which imagination could run free. There was little impetus for any sustained attention to the past, and relatively little was published that could be called historical.

The initial surge of interest did result in the 1911 publication in *The Scientific American* of Alfred Acton's long article.[2] In a much less schol-arly vein, four pages of the 1912 *The Boys' Book of Aeroplanes*[3] were devoted to Swedenborg's design.

World War I brought historical research to a virtual standstill. Aviation personnel found their energies absorbed in military projects, and Swedenborgian researchers were hindered by travel restrictions. It comes as a distinct surprise, then, to find an article on Swedenborg's design appearing in a German technical journal in the middle of the war.[4] Von Klinckowström treats the subject with characteristic German thoroughness, showing awareness of the Tafels' search for documents, the Odhner translations, and the 1910 Congress. His is distinctly more comprehensive than previous treatments, outlining the circumstances under which the design was made, Swedenborg's own reservations about it, and Polhem's objections.

ÅNGSTRÖM AND FOLLOWERS

Excluding passing references in various contexts (especially in Swedenborgian periodicals and bibliographies), the next significant attention given to Swedenborg's machine is by Tord Ångström in the 1920's. Ångström, later Inspector of Civil Aviation in Sweden, had a lively interest in aviation history.[5] As a young man he had studied flying under no less a personage than Louis Bleriot in Paris, becoming deeply involved in French discussions of the future of aviation. His profession would later involve him in the international cooperation in aeronautics which began with the Versailles Treaty and the Paris Convention on Civil Aviation of 1919; and this activity had the fringe benefit of conversations about early inventors.

When he published an article on civil aviation in *Upfinningarnas Bok* (*The Book of Inventions*) in 1920, he apparently was not aware of Swedenborg's design, since there is no mention of it in that work. However, when he revised his article for a second edition in 1925, he included the design with his own thoughtful comments. He emphasized the fixed wing concept with the separate means of propulsion, and contrasted this with the earlier reliance on wings flapped by arms or legs.

His interest, once aroused, was enduring. He published a more thorough study of da Vinci and Swedenborg in 1934, with even greater appreciation of the originality and genius of the latter's design.[6] He published still another article, somewhat expanded but not substantially changed, in 1960.[7]

In 1927, two distinguished aviation writers also published books with references to Swedenborg's invention. The French historian Jules Duhem, in a work devoted to the Montgolfier balloon, presented an evaluation of Swedenborg's machine and compared it with the creations of Burattini and Musgrave.[8] He returned to the subject some years later with an interesting presentation in a pictorial work.[9] Also in 1927, the English writer C.L.M. Brown published a work which included comments on the machine.[10]

GROWING INTEREST

Late in the 1930's, there seems to have been a new surge of interest. We may here take explicit note only of some of the more significant books and articles from this period, referring the reader to the Bibliography for a more complete listing.

The year 1938 was the two hundred and fiftieth anniversary of Swedenborg's birth, and a "post-London" generation of Swedenborgians devoted its energies to observing the occasion. The contribution of one of the "old-timers," the Rev. Alfred Acton, was a milestone work, *The Mechanical Inventions of Emanuel Swedenborg*.[11] He presented everything available from Swedenborg's hand, including of course the letter of September 8, 1714. He provided his own translations from Swedish and Latin where necessary, and modernized the Odhners' versions of the manuscript and the published account. His comments on the airplane do not break new ground, but are eminently fair, straightforward, and understandable.

Of particular interest in Acton's presentation is the first detailed technical realization of Swedenborg's rough sketch of the aircraft. The draftsman was Gustaf Genzlinger, a Swedenborgian who devoted a great deal of time to the mechanical works, and whom we have already met as the builder of the Smithsonian's model.[12]

On the more popular and public side, in celebration of Swedenborg's 250th anniversary, the January 1938 issue of *Popular Aviation* carried an article by Margaret Wales entitled "Emanuel Swedenborg, 18th Century Pioneer." Wales took due note of the airplane design. The magazine was widely circulated, and adaptations of the article appeared in several newspapers and magazines.[13]

The article had one other immediate effect. In 1930, an American writer on aviation history named John Goldstrom had published a book called *A Narrative Story of Aviation*. It contained a survey of early attempts to fly, ending with the success of the Wright brothers, with no mention of Swedenborg. In 1938, however, Goldstrom published an article in *The Swedish-American Monthly* with the title, "He Foresaw the Aircraft." By then he had discovered Swedenborg, and was bursting with enthusiasm. He wrote,[14]

...Swedenborg seems to have had no illusions that his flying machine was actually ready to fly, but he believed that he had indicated basic principles

43

for future development. Margaret Wales, an authority on Swedenborg, says, in "Popular Aviation" for January 1938, that these principles "have been pronounced sound, by modern aeronautical engineers".... At this point the present chronicler [Goldstrom], himself an aeronautical historian, will pause for a station announcement and admit his own comprehensive ignorance heretofore, of the Swedenborg design. How little has been known about it by specialists in my field may be indicated by the fact that in my own "A Narrative History of Aviation" I came across no reference to it in even the earliest aeronautical books.

Goldstrom further noted that a Dr. Victor O. Freeburg, in the Library of the American Museum of Natural History in Chicago, had discovered a facsimile edition of the first four issues of *Daedalus Hyperboreus*, "which Swedenborg edited and is said to have entirely written."[15] This provides us with a link to "the Glenview story," which we will present below.[16]

Goldstrom's article is essentially a condensation of the Odhner translations with Margaret Wales' general comments cited. It does not further our understanding of the design,[17] but it is worth citing as typical of the general neglect of the Swedenborg machine by aviation historians on the one hand and of the excitement of discovery on the other. Both sides of this situation are particularly vivid to the present author. While my involvement in aviation history is recent, my whole career has been in Swedish and Scandinavian aviation, and my obliviousness to Swedenborg's proposal was as complete as Goldstrom's—in spite of the fact that the manuscript and its drawing rested only a few hundred yards from my birthplace and childhood home![18]

GIBBS-SMITH AND OTHER HISTORIANS

World War II again interrupted normal academic travel and research. The publication in 1953 of Charles Gibbs-Smith's *A History of Flying*, however, opened many eyes to the fascinations of the prehistory of flight. Gibbs-Smith had started his writing career in 1942 with a work on aircraft recognition, a matter of particular importance during the war. He continued to update this with great skill, with new editions appearing almost yearly.

Then in 1948, he edited *The New Book on Flight*, a widely appreciated volume which surveyed the rapid development of aviation during the war. He continued his research and writing, and during the 1950's and 1960's became the world's leading authority on the history of aviation. In 1970, with the advantages of a twenty-five year perspective, he published

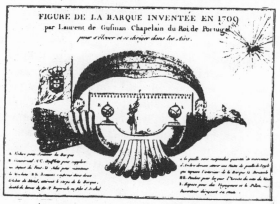

A reconstruction of Gusmao's "Passorola"

Aviation, A Historical Survey from its Origin to the End of World War II, which remains the classic treatment of the subject.

His evaluation of the design itself is reserved.[19] He does suggest that Swedenborg could have been influenced by Lourenço de Gusmão, which is not unlikely; and he believes that Melchior Bauer had in turn been influenced by Swedenborg.[20] More than other historians, then, he tends (rightly or wrongly) to see Swedenborg as a participant in and contributor to the progress toward flight.

The next wave of interest came in the 1960's, with the publication of Tord Ångström's *A Machine to Fly in the Air.* While it added little to his earlier treatment of the same subject,[21] it seems to have played a background role in prompting the construction of the model in the Smithsonian Air and Space Museum.[22]

Probably because of the high visibility of this model, (though note should be taken of one model in the Diocesan Library at Linköping, and another in the Stockholm Technical Museum)[23] references to the airplane became more frequent in newspapers and periodicals during the 1960's, and there is an increase in references to the subject in scholarly books.

Three of these deserve special mention. Gerhard Wissman's *Geschichte der Luftfahrt*[24] gives a sound and careful analysis, and also compares the machine to other projects. Wissman acknowledges his indebtedness to the 1916 account of his compatriot, von Klinckowström.[25] The Swiss Swedenborg scholar Ernst Benz, in his 1969 biography of Swedenborg, provides valuable references to his mechanical inventions, including the airplane.[26] Finally, Clive Hart's definitive *The Prehistory of Flight,*[27] published in 1985, gives the most comprehensive single treatment to date, dealing with literary and linguistic questions as well as with historical and technical ones.

AN "AIR CUSHION VEHICLE"

There is one intriguing footnote to this survey. In 1963, the English engineer Leslie H. Hayward wrote a book on air cushion vehicles.[28] He begins his exposition with the following sentence. "It may be of interest to know that the first proposal which I can trace for a machine which today would fall under some classified heading, subheading, or sub-subheading of ground effect machine was put forward by Emanuel Swedenborg, the Swedish scientist, philosopher and mystic during 1716." He then goes on to describe Swedenborg's early travels abroad, his interests, etc., leading up to a thorough description of the machine. He does not evaluate it explicitly, but does add that it was 160 years afterward that John Ward of San Francisco proposed a machine on the same principles. This thought has been further developed in a 1967 article in American periodical, *The Automobile Quarterly*.[29]

VIII

MODELS OF "THE MACHINE"

THE GLENVIEW EXPERIMENT

As indicated above,[1] Swedenborg's proposal was rescued from oblivion by individuals interested first of all in the man rather than in aviation. The institutionalization of this interest led to intensive searches for documents[2] and ultimately to significant collections of information, accessible to the general public, but hardly conspicuous.[3]

Swedenborgians were by no means immune to the general excitement about aviation, and there were individuals who took particular interest in the proposal for a flying machine. There is good evidence that a model of the craft was built and tested in the Chicago area in 1898 by a certain Jesse A. Burt, an employee of the Chicago Natural History Museum (later the Field Museum of Natural History). Since this was more than a decade before the Odhners' publication of the documents, his source must have been the facsimile edition of the first four issues of *Daedalus Hyperboreus* in the museum where he worked.[4]

Information about the experiment is scattered among archives at Glenview, Illinois, Bryn Athyn, Pennsylvania, Columbia University, New York, and the Technical Museum of Stockholm.[5] The primary sources consist of correspondence between interested Swedenborgians and relatives of Burt. Perhaps the earliest was written by Burt's sister to a niece:[6]

> Uncle Jesse's Glider was modeled after Swedenborg's plan from *Principia*, we suppose. One of his favorite books. It was pulled along by ropes until it ascended, and it was intended to be air propelled, and balanced by the movements of his body. The idea was rational, but he did not know enough about air currents. Fortunately, it was not a windy day, and he

came down close by watchers, still attached to the rope. In (Hammonds-port) No. Elmyra the air propelled Gliders fly often... Uncle Jesse came down in a side slip and was not hurt.... .

There is also an eyewitness account written in 1904 by William F. Junge, who had been a young boy when he saw the flight. His recollections are as follows:

THE FLYING MACHINE THAT WOULDN'T

An account given by the eye witness, W.F. Junge Feb. 1st 1904

Let the kind reader take a few minutes from his or her busy life and listen to the tale of a flying machine that wouldn't and he will be well repaid.

It was in the little village of Glenview that was destined to be the starting place of this great machine. On the 20th of May 1898, the inventor started from Chicago, at seven o'clock in the evening and arrived in Glenview three hours later with his machine which among its other qualities could be taken apart. The next morning, which, by the way, was Sunday, the inventor, with some of his friends, put the machines parts together. By noon the machine was ready for flight and by two oclock everything was ready in the way of ropes and other ground equipment.

Before I go farther on with my account it might be well to discribe the machine. It was built after Swedenborg's plan, in all but a few particulars one of which was the leaving out of a movable balancing weight. The machine had a large oval kite like surface, with a seat below for the navi-gator, for lifting and sailing this machine. The inventor expected to fly his machine like a kite until it got high in the air and then to cut loose and soar like a hawk. In theory his machine was all right, but that was all.

About half past two the inventor took his place and men manned the rope while others held the machine off the ground. The signal is given, and the men at the rope start running, and the machine slowly rises until it is about twenty feet from the ground. But when the machine was about twenty feet from the ground the inventor found that without a balancing weight he could not govern the machine and so, in spite of all he could do, the machine crashed to the ground.

When the machine landed a young man who was standing under it sud-denly found he had an enormous collar on. Fortunately no one was hurt and the only harm done was to the machine which was never to fly again.

The moral of this story I will leave out.

William F. Junge

Junge's father, W.H. Junge, wrote down his memories of the event nearly fifty years later.

BURT'S FLYING MACHINE

The notes below may not all be accurate, being based wholly on memories that have faded.

This much is true, the rig did fly under control for over a hundred feet, before the wind backing up under the peak sent it into a slip.

As I recall the fall may have been 50 feet but the wire [wing?] construction took the blow and broke the fall before breaking up itself. Jesse was not hurt in the least.

The contraption was got up in the air by means of a long light rope, perhaps a hundred feet long but there was only some 80 feet between the rig & some 15 men and boys running. It sailed up beautifully at the first trial. Jesse had cut loose and flown just a little.

Jesse crawled out of the wreckage and said "Well! I'll have to begin over again". I never heard of his doing so.

W. H. Junge, September 4th, 1946.

This statement is accompanied by a drawing of Burt's design, made from memory. There are similarities to Swedenborg's sketch, but also great simplifications: there is no way to know whether they are Burt's or Junge's. The cockpit in particular differs from those of later models, and may be more nearly what Swedenborg intended.[7]

The wave of interest in the airplane after World War II brought the Glenview story to the surface. There was a short story about it in a 1960 aviation calendar published in Sweden, and soon thereafter a Swiss Swedenborgian periodical carried a fuller account.[8] This article in turn was translated into English and published in *The Messenger*, an American Swedenborgian monthly. Finally, in 1962 a popular aviation publication in the United States drew on W. H. Junge's account in connection with a mention of Swedenborg's airplane.[9]

Kind members of the Swedenborgian community in Glenview are willing to show visitors the hill and the meadow where the experiment took place, and there are members of the Junge family to tell the stories they heard from their relatives of a generation ago.

THE 1910 LONDON MODEL

When the Plenary Session of the International Swedenborg Congress convened on July 5, 1910,[10] The President of the Congress, Edward J. Broadfield, included the following notice in his opening remarks:

I have been told that Mr. Rendell intends to produce some time during the Congress a model of this aeroplane. I am not in the confidence of the reverend gentleman as to this, but he will perhaps forgive me saying that if he intends to produce it as a practicable machine at the garden party this week, I hope he will not invite the President of this Congress to attempt the first flight![11]

The Rev. J.R. Rendell had undertaken to introduce three of Swedenborg's mechanical inventions to the Scientific Section when it met on the afternoon of the same day—the flying machine, a conveyor for ores, and an air pump. In his address to the meeting he summarized the principles Swedenborg incorporated in his design, illustrating his talk with projector and screen.[12]

The President's "fears" were unfounded. The model turned out to be in small scale. It has since disappeared[13], but it served its purpose of awakening interest.

A GERMAN MODEL

It is perhaps surprising that the next model we hear of is in Germany in 1916, in the middle of World War I. Count von Klinckowström illustrated his article on the machine[14] with two photographs of a model which he says was constructed in Munich by a Herr Amman. Unfortunately, nothing more is known about the genesis or about the fate of this model.[15]

The Linköping Model

Lars Ekelund, Linköping

MODELS IN SWEDEN

In connection with the celebration of Swedenborg's 250th birthday in 1938, a model was built and put on display in the Technical Museum of

Stockholm on the initiative of the Curator, Thorsten Althin. It was built to a scale of 1:40 by an employee and foreman of the Museum, Arvid Ericsson, a skilled model builder. It is on permanent display in the aircraft section of the main exhibition hall, with the following caption (in Swedish):

> SWEDENBORG. Arrangement for forward driving by the method of "air-oars" with coil springs placed below. For the first time the most important principles are indicated: a fixed surface with a seating box and a device for propulsion. Swedenborg himself said, "It seems easier to talk of such a machine than to put in into actuality and get it up into the air, for it requires greater force and less weight than exists in the human body."

Behind the Curator's initiative lay the persistence and the expertise of Tord Ångström.[16] Since 1925, he had been writing about the Swedenborg airplane without arousing much interest, but now the special occasion of the birthday provided a unique opportunity to catch the public fancy. In his 1939 article,[17] he maintained that the model makes it easier to understand what Swedenborg really had in mind: "even if the model looks rather awkward and funny compared with the more modern aeroplanes on display next to it. There exists a risk of underrating the value of Swedenborg's idea, something which is unjustified because it carries with it the most productive thoughts conceived until his time toward the solution of the problem of flight . . . ."

The Technical Museum has a second model, to a scale of 1:20, currently in storage. It was built in 1964 by Bertil Engman, and is used for exhibits outside the Museum. It was, for example, displayed at the Swedish Pavilion at the Trade Fair in Chicago in 1969.

Lastly, we may note that there is now a model in the "Curiosa Chamber" of the Diocesan Library in Linköping, diplayed together with the manuscript and the drawing. It was constructed by a local craftsman, Sigurd Molijn.

THE SMITHSONIAN MODEL

The model in the Smithsonian Museum has been seen by more people than any of the others. It is perhaps fitting that its genesis and the process of its construction are particularly well documented.

The initial suggestion was made by the Rev. Karl Alden, representing the Swedenborg Foundation in New York, to the Director and the Curator of the Air and Space Museum, Philip Hopkins and Dr. Paul Garber. Their interest was supported by Alfred Acton's thorough survey of Swedenborg's mechanical inventions[18] and by a review by Lennart Alfeldt[19] of an account of the machine by Tord Ångström.[20]

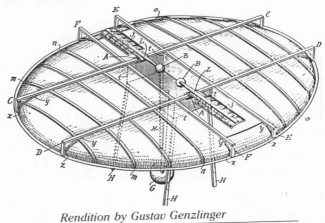

Rendition by Gustav Genzlinger

The Smithsonian has high standards of quality for its exhibits. The model-builder Gustav Genzlinger reports,

> To mention a few, they specified that; the oval frame, the bowed splints and supporting poles, as Swedenborg called them, be made of brass with all joints riveted and soldered; the sail cloth be of fine weave linen; the wings and other parts of seasoned hard wood; and the car of light-colored wood reinforced by splints, because Swedenborg suggested that it preferably should be made of birch bark with wooden pegs. They also specified a special kind of parallel stitching to hold the linen to the splints of the oval frame without stretching or wrinkling and without the use of glue or dope treatment of any kind. Also all exposed brass parts were to be stained to give the appearance of wood.
>
> The reason for the requirement of brass for the framework in this particular model is to provide a durable model that will retain its bowed shape and have lasting quality throughout the years.[21]

Genzlinger worked on the model all through 1961, with Thorsten Sigstedt carving the wooden parts and Mrs. Ariel Rosenquist doing the sewing. The Smithsonian kept a close eye on the process. Dr. Garber visited Bryn Athyn three times to inspect the model and examine its craftsmanship, and there was an exchange of some thirty letters during the period of construction. As an example of the kind of detail covered, we may quote from a letter from Garber to Genzlinger dated April 20, 1961:[22]

> The snapshots you enclosed, and my examination of the model reveal that you are indeed making excellent progress. The way you have constructed the air-beating surfaces (which Swedenborg called "wings") is a sensible interpretation of Swedenborg's drawing and description, and I agree that your method of attaching the louvre fabric along the central strips and the inner longitudinal edge, together with the addition of a

crosswise splint, will produce a workable valve-like action for opening the air-beating surface on the up stroke and closing it on the down stroke, thus imparting a downward and rearward pressure on the air. If we were making this aircraft full size we might find that the valve fabric, during manipulation, would tend to fold diagonally along the hypotenuse-line of the triangle bounded by the attached edges. That could be prevented by use of a hinge attached to the splint and the central spine. I am not suggesting that you try to hinge the splints. That would be virtually impossible at this scale, but, if you agree with my interpretation of the valve action, you might represent the hinge by a small piece of black sheet plastic glued to the splint and spine. I'll sketch this idea on back of this page and you can consider whether it makes sense or not. My thought is that Swedenborg, with his fine mind, would have realized that the fabric on the "wings" would have to be impermeable to air and yet sufficiently flexible to open and close. therefore he might have used a stiffener, such as starch, on the fabric. I also believe that glues were known at that time and might have been used to supplement the sewing and tacking. We not only have to think how we, as modelmakers, can represent the device, but also try to do some of Swedenborg's thinking for him as he would have proceeded to develop his invention.

Of course, we now know that the amount of down-and-backward air generated by these air-beating surfaces would have been far less than that effected by screw propellors, but that logic is based on 20th century thinking whereas you and I must consider this aircraft by dating our minds back to the early 1700's. Dating our appreciation from that time, this aircraft is certainly advanced over then-known ideas for flight. Da Vinci's wonderful concepts were at that time hidden away, and did not come to light until Napoleon's armies found them in Italy and brought them to France.

The way you have stained the metal, forms a good representation of wood. This effect will be enhanced when you apply oak stain to the brass-wood channel strips I am sending you. They can cap over the brass top-side girders. Have you tried any of the new epoxy resin glues? They make a good bond between wood and metal. The natural light color of the car is realistic.

The knobs on the handles seem to be of right proportions, and the way you have sprung the handles is ingenious and workable. In our discussion of the sewing we agreed that it will be logical to have the fabric on the under surface and bring the edges around, up, and over the rims, forming the "sail" from two halves, fore and aft of the car, making the first attachment in line with the longer dimension of the car, and with the "sail" detached from the car for easier handling. Then the fabric would be pulled taut to the end of the ellipse, and sewed around the curve, possibly using two needles alternately so that the attachment could proceed evenly both ways from the end curve, and avoid wrinkling. Spring clip clothespins would be useful to hold the fabric to the rims as the sewing proceeds. We agreed that inasmuch as Swedenborg specified linen for the "sail" we should try to get that fabric. Is it possible to wash the stiffener out of tracing cloth? If so that might be a good fabric. The sizing could be left in for the fabric used on the "wings". The weave should be in scale, and the color could be dyed a tan, buff, or light brown. Maybe a diluted dye would produce the right color.

The Genzlinger model under construction,
showing the cockpit, handles, and springs

As a further example of the close attention to detail, we may turn to the question of the orientation of the pilot's car, and cite Genzlinger's own reasoning:[23]

Reverting now to the construction of the pilot's car in the present model, which I have referred to as being wider than it is long, some explanation is necessary because I feel quite sure that this will be questioned by someone—particularly since a photograph, published in the "New Church Messenger" of May 1960, of the model on exhibit in the Technical Museum in Stockholm shows the car as being longer than it is wide, just the opposite of the present model.

Twenty-three years ago, when I made the reconstruction published in the "Mechanical Inventions of Swedenborg", I pointed out to Dr. Acton that the paragraph describing the car had a contradiction in it. It read "The box should be two ells from back to front and three in breadth, inasmuch as the wings must be moved at the sides", and then goes on to read "this (referring back to the breadth) therefore should be smaller than the length". In making the reconstruction drawing of that publication I followed the two and three ell dimensions and therefore that drawing is the same as the present model.

Even though Swedenborg's sketch appears to show the long dimension running lengthwise I felt justified in making it crosswise because of his statement, "three ells in breadth, inasmuch as the wings must be moved at the sides". This meant to me that more space was needed crosswise in order to have the arms of the wings project for a sufficient distance into the car to get adequate leverage for their operation with minimum effort.

While I was working on the model, Miss Beryl Briscoe supplied me with a copy of Carl Th. Odhner's translation published in 1910, which I did not have when making the reconstruction drawing, and I noted that his translation used the work "greater" at the point in question intead of the world "smaller". You can imagine how pleased I was to find this. And now I find an article by Lennart Alfeldt in the October-December 1961 issue of the "New Philosophy" in which he calls attention to an error in the first-mentioned translation in that "smaller" should have read "broader".

I have dwelt rather fully on this seemingly small point because Mr. Garber noticed it and after discussing it with me agreed with my conclusion, and because it came up again when I displayed the partially completed model at a Swedenborg Scientific Association meeting, and finally in Lennart Alfeldt's article. This shorter than wide or wider than long has bothered me for some time, but now, thanks to Dr. Odhner's translation and to Lennart Alfeldt's article, I can sleep in peace in my bed which is "longer than it is wide—either crosswise or lengthwise".

Bearing in mind that Garber's was just one letter out of many, the reader may feel assured of the authenticity of Genzlinger's model.

THE BRYN ATHYN MODEL

A replica of the Genzlinger model is kept in the archives of the Sweden-borg Scientific Association in Bryn Athyn, Pennsylvania. It differs slightly in some details, in hardware materials, and in fabrics, and for educational purposes has explanatory labels attached to the various parts.

The Bryn Athyn Model

The Australian Model

This is obviously valuable to the student, but somewhat disturbing visually, obscuring the essential economy of Swedenborg's design.

THE AUSTRALIAN MODELS

An Australian electrical engineer and amateur model-builder, A.G. Taylor, has constructed a copy of the Genzlinger model. When not used in connection with lectures and seminars, it is on display in the Swedenborg Center, Roseville, N.S.W., outside Sydney.

As this chapter is being completed, word also comes that in connection with the coming tricentennial of Swedenborg's birth, a model is being produced in Australia through the cooperative efforts of Roland Smith of Kent, England, and Neville Jarvis of Roseville, Australia. It will be made of light cardboard, simplified so that young people will be able to cut out the parts and assemble them. Like the original machine, it comes without any guarantee that it will actually fly!

THE PRESENTATION OF THE SMITHSONIAN MODEL

There is no fully appropriate way to end a story that is still going on, but we may look at two significant summaries of the subject.

The first centers in the formal presentation of the Genzlinger model to the Smithsonian Museum. It was a memorable occasion, held in the board room of the New York headquarters of the Scandinavian Airlines System, with distinguished representatives of the various coinciding interests. Aviation was represented by the Scandinavian Airlines System,

The Genzlinger model in the Smithsonian Museum

Sweden by the Swedish Embassy, and Swedenborgians by the Sweden-borg Foundation; and above all, Mr. Genzlinger himself was in atten-dance. The Rev. Karl Alden presented the model to Dr. Paul Garber of the Smithsonian.

The model was displayed in the street windows of the SAS offices in New York and Washington, D.C., until January 29th, 1962, Swedenborg's 274th birthday, when it was placed for permanent exhibition in its pres-ent location. On this occasion, the Swedish Ambassador, Dr. Gunnar Jar-ring,[24] gave an extensive interview over WRC's "Capitol By-Lines," interviewed by Patty Cavin. We may quote from the close of the inteview.

Ambassador Jarring: Swedenborg's inventions in different technical fields can be counted in the hundreds. Most of them were never put to test, though, as he disliked experimental work, in spite of the fact that he was himself skilled in many trades, like book-binding, clock-making, engraving and lens-grinding. I have already mentioned his air pump. He made improvements in the primitive hearing aids of the day—the ear trumpets. He made a musical machine, forerunner of our phonograph. He built an experimental tank for testing ships, similar to those which are still in use. He reflected on the possibility of a submarine, designed a machine gun, and marketed a usable fire extinguisher. And he designed a glider-type aeroplane, which was the reason that brought us here today.

Mrs. Cavin: Will it fly, Mr. Ambassador? Would you care for the job of test pilot?

Ambassador Jarring: Well, I am not too sure about that, and I doubt that Swedenborg himself considered his "Machine to fly in the air," as he called it, to be among his most important inventions, or that it could become of any particular blessing to humanity. Still, it has been called the "first rational design for a flying machine of the airplane type" by such a distin-guished authority as the Royal Aeronautical Society in London.

Mrs. Cavin: Thank you, Mr. Ambassador. I think you have given us a pic-ture of a most unusual man. I think, though, that to most Americans, Swedenborg is known as a religious teacher, as the founder of one of our innumerable church groupings.

Ambassador Jarring: Well, this is also true, and this is the other and no less fascinating aspect of the man Swedenborg. He was a true child of the Age of Reason. "The world is nothing but a machine," he wrote in his early years; the soul, human life and character, are nothing but tremulations in the body's material particles. So he set out searching for the soul by studying the mechanism of the human body. From his fifty-seventh year to his death he carried on this search, moving with every year farther from the clearcut rationalism of his early years, into the mystic depths of spiritual life. His religious works, which cover more than 16,000 pages, have been translated into more than twenty languages, including the Braille alphabet for the blind, which opened the world of Swedenborg, for instance to people like Helen Keller. He has between five and six thousand followers in this country who gather to study his teachings and ideas in some fifty churches all over the country, of which the famous Wayfarers Chapel in Portuguese Bend, California. . . is the best known.

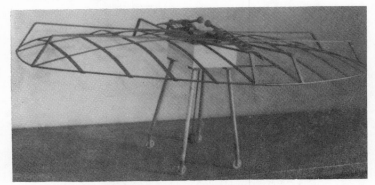

Framework of the Genzlinger model

A SUMMARY EVALUATION

We may close by citing Genzlinger's list of the features of the design which were of greatest interest to the experts at the Smithsonian. They were:

1. Swedenborg's realization that adequate sail surface (45 feet × 60 feet) would be required to support the weight of a man and an integral propelling apparatus, and his provision of a sail curved both longitudinally and laterally. In this respect the sail is similar to the wing used by the Wright brothers in the successful Kitty Hawk flyer, which wing was curved downward fore and aft and also from end to end, with a spread of over 40 feet.

2. The louvre-like valves in the paddle wings and their backward tilt to give a forward impulse.

3. The coil-spring mounting of the paddle wings to augment the pilot's ability to operate them and to increase the vigor of the downward or propelling stroke.

4. The undercarriage, a necessary accessory, that was suggested by Swedenborg over 100 years before the first illustration of wheeled craft found in our research.

5. The use of a suspended weight to maintain balance.

Thus Swedenborg, as early as 1714, suggested unique features such as wing, manual control, and structure which eventually appeared in realizations of successful flight.

APPENDIX

SWEDENBORG'S CONTACTS ABROAD RELATED TO HIS MECHANICAL INTERESTS

Both Bishop Swedberg, Swedenborg's father, and Eric Benzelius, his brother-in-law, had travelled abroad and had established high reputations in professional circles. Their letters of recommendation for young Emanuel therefore carried considerable weight, and his fledgling association with Christopher Polhem could have done no harm. Through the Swedish Ministers and the personnel of the legations in London, the Hague, and Paris, many doors were opened to him which would have been closed to students with less influential sponsors.[1]

With his strong personal ambition and apparent complete self-confidence, Emanuel used these credentials to the fullest, not simply for his own sake, but also for the benefit of his principals in Sweden, primarily the Collegium Curiosorum.[2] For obvious reasons he took a special interest in meeting great men in his own favorite fields—mathematics, algebra, geometry, and astronomy—this last being a subject which fascinated him more and more.[3] His contacts provide an index of his interests.

John Flamsteed (1646-1719) was the first Astronomer Royal at the Greenwich Observatory. His *Historia Coelestis*, the first of the Greenwich star catalogues, was published during Swedenborg's stay in England. Swedenborg's discussions with him focused on the motion of the moon, with the ultimate objective of using it to determine longitude.[4]

Edmund Halley (1656-1742), English astronomer and mathematician, is historically famous for his prediction of the return of the comet that now bears his name, and was the first to discover the use of a transit of

Venus in determining the parallax of the sun, a subject which fascinated Swedenborg. In 1676, he had published a catalogue of 341 stars in the southern hemisphere. He had also been instrumental in the publication of Newton's *Principia*. Emanuel discussed the longitude problem with him at some length.

John Woodward (1667?-1728), clergyman and mathematician, became Newton's successor as professor of mathematics at Cambridge in 1703. He was expelled from this post in 1710 because of his Arianism, a theological stance in opposition to that of the established church. In 1712, he announced in the *London Guardian* that he had a new proposal on the longitude problem, which prompted Swedenborg to accelerate his own studies in order to meet a new challenge. Swedenborg studied his *A New Theory of the Earth* (1696), *Praelectiones Astronomicae* (1707), and *Praelectiones Physico-Mathematicae*, and also met him and discussed mathematical problems with him.

Francis Hauksbee (1687-1763) was a renowned maker of air-pumps, telescopes, and other precision instruments in London, and also a fellow of the Royal Society. Swedenborg met him and procured both a description and a drawing of his air pump. There seems to have been more than a business relationship between the two, who were of much the same age. Swedenborg's intense interest in mechanics made Hauksbee's skills particularly attractive. In 1709, Hauksbee had published his *Physico-mechanical Experiments*, which Swedenborg bought for the Library of Upsala University. It was a small work, but full of novel experiments which opened entirely new fields of investigation. Some of the inventions which Swedenborg mentioned in his letter to Benzelius from Rostock may have had their origins in this book.

Baron Johan Palmqvist (1652-1716) was the Swedish Envoy to the Netherlands, and at the time of Swedenborg's visit, Sweden's representative at the Utrecht Peace Congress. The two shared an interest in mathematics, and the Baron gladly gave the young man abundant time to discuss their common interest. They also discussed Swedenborg's proposal for the founding of a Swedish society for physics and mechanics like the Royal Society in England, an idea close to Swedenborg's heart for many years.[5] Palmqvist and other members of the legation introduced Emanuel to many scholars, and enabled him to visit the observatory at Leiden.

Abbé Jean Bignon (1662-1743) is generally regarded as one of the most learned men in France at his time—"the Maecenas of his age and the guardian angel of the sciences and learning."[6] Swedenborg's letters from Palmqvist and Benzelius assured him of a cordial reception. Bignon was known especially for his culture, for the immense scope of his reading, and for the encouragement he gave to others, in this respect resembling Eric Benzelius. He was the editor of the *Journal des Sçavans*,[7] and was appointed Royal Librarian in 1718.

Paul Varignon (1654-1722), to whom Emanuel was introduced by Abbé Bignon, was a member of the Royal Societies of both France and England, Professor of Mathematics in the Mazarin College, and Professor of Philosophy at the Royal College. Swedenborg had long discussions with him about his inventions. He seems to have made a very favorable impression, since he visited him several times in the course of a single week.

Phillipe de la Hire (1640-1718), the great French astronomer, was an intimate friend of Varignon, who doubtless introduced Swedenborg to him. They discussed the problem of longitude, and de la Hire was surely interested in the young Swede who was fresh from discussions with Edmund Halley.

Father le Quien (1661-1733), a friend of Benzelius, was the Librarian of the Convent of St. Germain. He was delighted to show Emanuel around and to introduce him to libraries, books, and items of cultural interest.

Father le Long, another friend of Benzelius, was Librarian of the Oratory of Paris. He introduced Swedenborg to his work in progress, *Bibliothèque Historique* (1719), which listed all known books and manuscripts in the history of France.

There is every reason to assume that Swedenborg's ideas on many subjects, including the airplane, were developed and refined by his contacts with these most able minds.

NOTES

I   DREAMS OF FLIGHT

1. On December 17, 1903, at Kitty Hawk, North Carolina, Wilbur and Orville Wright made the first recorded flight in a heavier-than-air craft that met the criteria of being powered, self-sustaining, and controlled. Both machine and engine were of their own construction. The flight lasted 59 seconds and covered 852 feet.

2. Countless numbers of books have been written about the flight of birds. In addition to Leonardo's, some of the most significant historically are those of Roger Bacon (1250), Pierre Belon (1551), Hieronymus Fabricius (1618), Giovanni Borelli (1680), Chevalier de Vivens (1742), Johan Silberschlag (1871), Francois Huber (1784), Paul-Joseph Barthez (1798), Louis-Pierre Mouillard (1881), and Otto Lilienthal (1889). For more detailed information, see Clive Hart's *Prehistory of Flight,* Part I, pp. 1-115, and Charles H. Gibbs-Smith, *A History of Flying* (1953), Chapter II, pp. 28-52. Swedenborg also studied the flight of birds and made references thereto in his manuscript on the "machine"; see pp. 19 and 22 *supra.*

3. "Ornithopter," a heavier-than-air craft deriving both support and propulsion from flapping wings in imitation of the flight of birds. No successful machine of this type has been produced. *Columbia Encyclopedia,* 2nd Edition, *s.v.*

4. Manpowered flights with lightweight carbon fiber material have become possible only in very recent years, The greatest achievement at this writing is the crossing of the English Channel on June 12, 1979, by Allan Brian of England. His craft, "The Albatross," weighed only 27.2 kilograms, and the flight lasted two hours and forty-nine minutes. The plane is now on exhibit in the Musée de l'Air at Le Bourget airfield, near Paris. On May 11, 1984, a similar flight was made at Hanscom Field, Massachusetts. See further Keith Sherwin, *Manpowered Flight* (Kings Langley, Hurts; 1975).

5. *cf.* pp. 4ff. and 29 *supra.*

6. "Leonardo da Vinci, Aviateur," by Abel Hureau de Villeneuve, in *L'aéronaute,* 7e Année, No. 9, Sept 1874.

7. Charles H. Gibbs-Smith, *op. cit.*, p. 46.

8. This historic first ascent took place at the Castle la Muette in the Bois de Boulogne, Paris, lasting twenty-five minutes and covering five and a half miles. There is a striking similarity between the Montgolfier brothers Joseph (1745-1810) and Étienne (1745-1799) and the Wright brothers (see pp. 30 *supra*). The first flight with a hydrogen balloon, constructed by J. A. C. Charles, took place on December first of the same year from the Tuileries Gardens in Paris.

9. Literature on this subject is voluminous. See Clive Hart, *op. cit.* (Bibliography), pp. 227-274.

10. The Library of the Smithsonian National Air and Space Museum lists approximately seven hundred titles in the category "Aviation History."

11. A steam engine for practical use was invented by the Scotsman James Watt in 1769. Swedenborg heard about experiments with steam power for pumping water in coal mines around 1720, six years after his first airplane design. Hydrogen gas for lighter-than-air balloons was first used successfully in 1783, some eleven years after Swedenborg's death. (See Note 8 *supra*).

12. Clive Hart, *The Prehistory of Flight* (London, 1985). Hart, Chairman of the Department of Literature at the University of Exeter, England, is also the author of *Kites: An Historical Survey* (1967) and *The Dream of Flight* (1972). See Bibliography.

13. Several of these individuals are treated also by such authors as John Alexander (*The Conquest of the Air*, 1902), Wilhelm Balthasar (*Die Anfänge der Luftfahrt*, 1909), C. L. M. Brown (*The Conquest of the Air*, 1927) and Charles H. Gibbs-Smith (*Aviation: An Historical Survey*, 1970). See Bibliography.

14. Swedenborg was certainly familiar with the works of the Franciscan monk Roger Bacon (1214-1292), whose name occurs in the listing of Swedenborg's personal library (Archives, Bryn Athyn Academy). Bacon, "the admirable doctor," was an outstanding scholar in many fields. In *The Philosophical Transactions, Abridged to 1700*, (London, 1705), J. Lowthorp cites Bacon as having said that "he himself knew how to make an engine in which a man sitting might be able to carry himself through the air like a bird" and that there was a man "who actually tried it with good success." There is no further evidence of this: Gibbs-Smith (*op. cit.*, p. 40) finds it difficult to accord this admittedly great man much credit for his aeronautical remarks.

15. Swedenborg, who was well informed in many fields, had surely heard of Leonardo as an artist. However, he can have had no knowledge whatsoever about his designs in the mechanical field, since these became known in Europe only during the nineteenth century (*cf.* p. 2 *supra*).

16. Clive Hart (*op. cit.*, pp. 135-45) recognizes Burattini as one of the most spectacular of the seventeenth-century experimenters. "However inadequate the principles he applied, Burattini worked with real seriousness and completed a rational design for what he believed would be a fully controllable ornithopter. Although of much lower intellectual and imaginative calibre than Leonardo, Burattini achieved the distinction of building and testing, reputedly with success, the first scale model of a flying machine to carry men." The general neglect of Burattini among prehistorians of aviation must be regarded as unjust. There

are no indicatons that Swedenborg knew about Burattini and his model. We may, incidentally, note a reference to him in the *Histoire Comique des États et Empires de la Lune* (1648), pp. 148f., of Cyrano de Bergerac (1619-1655).

17. Swedenborg was definitely familiar with de Lana's creation, which he scornfully, indirectly, and perhaps unjustly dismissed as "impracticable" and as "contrary to our atmosphere." See his remarks in the published account (*Daedalus Hyperboreus* No. IV), p. 21 *supra*.

18. The *Journal des Scavans* is included in the listing of Swedenborg's personal library (Archives, Bryn Athyn Academy).

19. Carl von Klinckowström has dealt with this subject in *Archiv für die Geschichte der Naturwissenschaften und der Technik* (1911), pp. 21ff. Another German author, J. Feldkirch, published an extensive article on Gusmão in the *Illustrierte Aeronautische Mitteilungen* (August 25, 1909), asserting without hesitation that Gusmão made the first successful trial of a hot air balloon in 1709, 74 years before the Montgolfiers'. There is no direct evidence that Swedenborg saw the engravings or descriptions of Gusmão's "Passarola." However, it must be assumed that he was familiar with the experiments, since they were discussed throughout Europe. The London newspaper *The Evening Post* of December 20 and 24, 1709 (Nos. 55 and 56) reported the experiment (it may be found in the Nichols Collection of the Bodleian Library at Oxford, England). Anyone in England interested in the problems of flight must have seen this article, and there is therefore every reason to believe that the insistently inquiring Swedenborg would have become aware of it.

20. The technical features of the craft, and their historical importance, are treated at greater length below in Chapters Five and Seven.

21. See pp. 32f. *supra*.

II   BACKGROUND

1. In the preface of his book on Swedenborg's *Journal of Dreams* (*Drömboken*, 1964), the noted Swedish author Per Erik Wahlund writes, "The combination of such delicate elements as religious mysticism and mental illness is perhaps what in the first place made Swedenborg a stranger to his own countrymen.... It is amazing that his position as a versatile genius should remain so unknown. While his production... is accessible to the English public in first class editions, it remains with us [the Swedes] on bookshelves, collecting dust in old fashioned and incomplete translations, hobbled by their Latinate sentence structure...."

2. "I was first introduced by the Lord into the natural sciences and was thus prepared; and this from the year 1710 to 1744, when heaven was opened to me." From a letter to Oetinger, as cited in R. L. Tafel, *Documents concerning the Life and Character of Emanuel Swedenborg*, Vol. II, p. 275.

3. One of the earliest of Swedenborg's biographers offers the following view of young Emanuel: "Religion is not any more in his thoughts. He has left the angels of his childhood, not in contempt, but in forgetfulness, having other business on hand. Of reverence he has plainly little, of self-satisfaction, much." (William White, *Emanuel Swedenborg* [London, 1868], p. 49.)

4. "Swedenborg lived at a time when experimental philosophy, as science was then called, had just established its hold on men's attention. From reasoning about the world in the study, men turned to examining it in the observatory and the laboratory. This transition from the rational to the empirical, like all fundamental revolutions, was not completed in a moment..." (Extract from an address, "Swedenborg as a Physical Scientist," given by Herbert Dingle, D. Sc., A.R.C.S., at Swedenborg's 250th birthday celebration in Queen's Hall, London, January 1, 1938).

5. *Christina* (1626-89), Queen of Sweden, was the daughter and successor of King Gustavus Adolphus II, who defended Protestantism against Catholicism in Europe during the Thirty Years' War. Her early devotion to affairs of state soon gave place to other interests. She attracted many artists and scholars, notably Descartes, to her court. At the end of her reign, she became attracted to Catholicism and abdicated in 1654. She spent most of the latter part of her life in Rome, where she died, being buried in the crypt of St. Peter's.

6. Afflicted by a lung disease, Descartes was unable to endure the rigors of the northern climate and died in 1650, a year after his arrival in Sweden.

7. In Swedenborg's early letters, written soon after he had left the university, he mentions the following as some of his professors in the Faculty of Philosophy: Professor of Mathematics Harold Wallerius, Professor of Astronomy Pehr Elfvius, and Professor of Elocution Johannes Upmark (later ennobled with the name Rosenadler). He also praises Professor of Theoretical Philosophy Fabianus Törner, who presided over Swedenborg's disputation in 1709. Törner's lectures comparing Aristotle's Logic to that of Descartes made a profound impression on young Swedenborg. See Claes Annerstedt, *Upsala Universitets Historia, 1877-1914*, Vol. II.

8. *Charles XII* (1681-1718), king of Sweden 1697-1718. In 1699, at the age of seventeen, he opposed a coalition of Peter I of Russia, August II of Poland and Saxony, and Frederick IV of Denmark, defeating Frederick and Peter in 1700 and August in 1702. In 1705 and 1706, he was at the height of his political and military power. From 1708 on, one disastrous defeat followed another, 1709 being the most notable turning point. On the relationship between Swedenborg and Charles see pp. 24ff. *supra*.

9. *cf.* pp. 8f. *supra*.

10. In June 1709 he "submitted to public examination" in the Gustavian Auditorium a thesis on selected sentences from the sayings of Publius Syrus Mimus, a popular Roman author of short plays. His father, Jesper Swedberg, by then a bishop, took part in the "disputation," which marked the close of Emanuel's university days.

11. *cf.* p. 9 *supra*.

12. The first Duke of Marlborough, John Churchill (1650-1722), is recognized as one of the greatest military commanders in history.

13. Hans Helander, in his excellent edition of Swedenborg's *Festivus Applausus in Caroli XII in Pomeraniam suam adventum* (Uppsala, 1985), deals with "Swedenborg's Attitude toward Charles XII" on pp. 18-20. It is clear that while Swedenborg remained outwardly loyal, he harbored strong inner reservations about Sweden's chances for recovery under Charles' leadership.

14. In his later years, Swedenborg would see the king as governed by an insatiable megalomania, using such epithets as "hypocrite," "liar," and even "mad," "…the most obstinate and stubborn man in the world… ," "…who refused to return home and make peace; he wanted to be the greatest of them all…." See A. Fryxell, *Berättelser ur Svenska Historien*, Part 21, pp. 209-211; also E. Swedenborg, *Diarium Spirituale* III, nn. 4741-48, 4758.

15. This was one of the last victories. The king himself was still in Turkey when the battle took place at Hälsingborg on February 28, 1710. The victor and hero was field marshal Magnus Stenbock.

16. He left in a great hurry, without saying goodbye even to his friend and mentor Eric Benzelius, just when the latter had made arrangements for him to stay with Christopher Polhem at his manufacturing plants at Stjärnsund. He apologized for his hasty departure in a letter of October 23, 1710. One factor in his haste may have been a favorable opportunity for passage, but Swedenborg may also have feared that involvemment with Polhem would preclude his foreign journey. Then too, there was widespread panic over the plague, especially at Upsala.

17. In London, he tells of visiting the shops of booksellers and instrument makers and gathering a "small stock of books on Mathesis" as well as various instruments such as a tube, quadrants, prisms, microscopes, scales, and a camera obscura. Among his books were Vitalis, the *Lexicon Mathesis*, and Newton's *Principia*. He worked at his English and read Newton "daily" in Latin. The *Principia* had been published in 1685, but the first known references at Upsala to his law of gravitation are in disputations in 1703 and 1716 under Professor Elfvius. The latter is skeptical about Newton's discovery as a "pure abstraction." Indeed, according to the *Upsala Universitets Historia* (pp. 323f.), "it was first through Swedenborg's visit to England, his correspondence and work, that the new views came to be adopted" in Sweden. See also S. Rydberg, *Svenska Studieresor till England under frihetstiden* (Upsala 1951). There are indications that Polhem did not read Newton until 1712, and then found his theories very obscure. See *Svenska Teknologföreningens Historia*, "Christopher Polhem" (Stockholm 1911), p. 62.

18. *cf.* pp. 8f. *supra*.

19. There is evidence that while in London during the last years of his life, when he had long since been primarily absorbed in the non-material world, he still worked at marble inlay as a hobby. He had developed the skill during a visit to Amsterdam in 1739. See *Svenska Linné-Sällskapets Årsskrift* 14, 1931, pp. 350ff.

20. He bought many rare books for himself, and his library over the years became a valuable one (See List, Archives of Bryn Athyn Academy). He was commissioned by his brother-in-law to purchase books for the Collegium Curiosorum, but he was also scouting for books, and made recommendations of his own. Some of the books in which he seemed to have taken special interest can be mentioned, especially since he may have drawn inspiration from some of them for his thoughts about the flying machine: Norris, *Reflections upon the Conduct of Human Life*; Baker, *Reflexions upon Learning*; Gregory, *Astronomia*; Halley, *A Collection of some Natural Phenomena*; "History of the Learned" (a

monthly book review); Hauksbee, *A Cathalogue of an Improved Air Pump* and *Physico-Mechanical Experiments*; Wilkins, *Mathematical Works* and *Natural Magic*; Lawthorp, *Magnae Britanniae Notitia*; and Harris, *Lexicon of the Sciences and Arts*. See Alfred Acton, *Letters and Memorials* (Bryn Athyn, 1948-55), pp. 18-56.

21. In his preface to "Suggestions for a Flying Machine," (1910, translation by H. Lj. and C. Th. Odhner), Alfred Acton proposes that the idea might have been conceived in 1711 or 1712, but cites no clear evidence. Carl von Klinckowström, in "Emanuel Swedenborg und das Flugproblem" (1916), shares Acton's belief.

22. See Alfred Acton, *Letters and Memorials*, pp. 51, 54.

23. *cf.* pp. 4f. *supra*.

24. *cf.* pp. 5f. *supra*.

25. "Evening Post," London, December 20-22, 1709, No. 56. See p. 65 *supra*.

26. *ibid.* pp. 218f. See also pp. 32 *supra*.

27. On October 7, 1772, the Director of the Swedish Royal Academy of Science, Samuel Sandel, delivered an obituary speech on Swedenborg to the Academy. In his remarks he laid particular emphasis on Swedenborg's status as an independent and original inventor and man of science.

28. A glead is a bird of prey, specifically a kind of kite, now quite rare, but in Swedenborg's time a common London street scavenger.

29. Kircher has been referred to as the inventor of the kite; but kites had been invented many centuries earlier. He devoted the last years of his life to the study of Egyptian hieroglyphics.

30. See pp. 4f. *supra*.

31. See pp. 69f. *infra*.

32. On Swedenborg's later relationship with Polhem, see pp. 25f. *supra*.

33. See pp. 21f. *supra*.

34. See p. 2 *supra*.

35. See pp. 15f. *supra*.

III THE MACHINE: "MANUSCRIPT" AND "PUBLISHED ACCOUNT"

1. See p. 37 *supra*.

2. Earlier, Swedenborg's intense interest in things mechanical seems to have focused on Polhem's inventions. In an earlier letter to Professor Elfvius, dated August 15, 1712, he had written about a forthcoming list of inventions. In the letter of September 8, 1714 to Benzelius from Rostock he wrote, "I have now a very great desire to go home to Sweden and take all Polhammar's [= Polhem's] inventions in hand, making drawings of them and giving descriptions; also comparing them with physics, mechanics, hydrostatics, and hydraulics, and likewise with algebraic calculations; and then publishing them in Sweden rather than elsewhere, and to establish a Society in Mathesis for us, for which we have such a fine basis in Polhammar's inventions. I wish that mine could serve this purpose as well." (Translation adapted from that of Alfred Acton, *Letters and Memorials*, pp. 58f.)

3. Rostock was in Pomerania in North Germany, now part of the German

Democratic Republic (DDR). In Swedenborg's time it was a province of Sweden, and it remained so, with some interruptions, until 1807.

4. See Chapter VIII.

5. Greifswald is also in Pomerania, about fifty miles east of Rostock, and is the site of a university founded in 1456.

6. The Odhners date the manuscript to September 1714. Swedenborg's letter to Benzelius, however, does not indicate which drawings are completed and which are still in process. We also have no way of knowing whether it was sent directly to Benzelius, or was, like other documents, forwarded by Swedenborg's father. In this latter connection, it may be mentioned that Emanuel needed his father's financial support, making it prudent to convince him that he was using his time productively. At this time, Jesper was Bishop of Skara, with his residence at Brunsbo, an official residential estate near Skara. In 1712, a fire destroyed the house and in all probability letters and documents from Emanuel.

7. I am indebted to Birger Holmer, retired Vice President for Aircraft Research and Development of the Scandinavian Airlines System, for examining the original texts and providing me with observations on and corrections to the Odhner translations.

8. "Ell" = Swedish *aln*, a unit of length. The Swedish *aln* used in Swedenborg's time = 0.60 meters or two feet; the old English ell = 1.14 meters or 1¼ yards.

9. Swedenborg used the Swedish *segel*, "sail," in paragraphs 2 and 3 for what today would be called a "wing," a fixed surface. He used the Swedish *vinge*, "wing," for what more correctly might be described as an "arm" for beating up and down. Some historians have referred to the latter as an "oar," which is incorrect, since these arms were designed to be operated up and down, more like a windmill than like an ornithopter. They definitely were not designed for rowing.

10. *sling fjaeder*, here correctly rendered "spiral spring," occurs as *sting fjaeder* in the published account. This must be a typographical error, perhaps due to Swedenborg's handwriting, since there is no such thing as *sting fjaeder*.

11. These "rollers" are not shown in the accompanying sketch.

12. For "Kirchberg," read "Kircher." The mistake is due to Swedenborg's handwriting. The Odhners were probably not familiar with Kircher as a resource for Swedenborg in the development of his design.

13. The published account locates the event at Strängnäs. Skara, as the site of Bishop Swedberg's church, may be correct.

14. Christopher Polhem (1661-1751), whose family name was Polhammar before he was ennobled, was also known as Polheim and Polheimer. The family was originally from Austria. His mechanical genius was discovered early, when he manufactured the ingenious clock in the Cathedral at Upsala. He was appointed to the College of Mines, where he soon rose to eminence because of his superior qualifications and his practical inventions. After an extensive tour abroad in 1694-6, he spent sixteen years as chief of the mechanical installations at the copper mine of Falun, then the top industry of Sweden. In addition to meeting his extensive official responsibilities, he instituted a mechanical laboratory and his own mechanical workshops at Stjärnsund in southern Dalecarlia,

in the mining district. His influence was enormous, especially since Sweden was then at war with powerful neighbors. His inventions covered a wide range, from clocks and locks to water dams and water locks, fortifications, and canals. As a son of the Age of Enlightenment, he also devoted time to writing on subjects like mathematics, physics, and cosmology, and even on medicine, history, national economics, the education of youth, religion, and languages. Swedenborg saw Polhem as the ideal mentor and sponsor, and Polhem was quick to recognize the young man's exceptional intelligence. Their relationship lasted for thirty-five years. It was indeed Polhem who introduced Swedenborg to the king, and the triangular relationship that developed became a cornerstone in the history of Swedish science and technology. Polhem, assisted by Swedenborg, designed and built dry docks and fortifications. Together, they drew up the original plans for the Göta Canal, traversing Sweden to link the North Sea with the Baltic.

15. Compared with most young men about to begin their professional careers, Swedenborg was wealthy. He now had an income from his share of the profits of the many iron works left to her children by his mother, Sara Behm, who had died in 1696, when Swedenborg was eight years old.

16. The first article in this issue of *Daedalus Hyperboreus* was written by Christopher Polhem, the primary contributor through the brief history of the publication. Swedenborg's flattering and exaggerated references to Polhem in other parts of the journal— probably part courtesy and part blandishment— may have been intended to blunt his objections to the flying machine. See pp. 59 *supra* and 71 *infra*.

17. This is clearly a reference to the design advocated by Francesco de Lana; *cf.* pp. 4f. *supra*.

18. See Note 13 *supra*.

19. See p. 13 *supra*.

20. Swedenborg had received a letter from Polhem dated September 5, 1716, in which the latter gave comments on the proposed contents of *Daedalus* No. IV. Swedenborg published the article in spite of Polhem's objections, indicating the strength of his belief in its merit. See also pp. 25 and 61f. *supra*. Polhem's objections may also have been the reason Swedenborg signed his article "N.N." rather than using his full name, perhaps suggesting his own strong ideas on the subject.

21. These are the last of Swedenborg's known comments about the flying machine. Several authors have toyed with the thought that discussion of the concept continued, both between Swedenborg and Polhem and between the two and Charles XII, but there is no evidence for this.

22. See. p. 26 *supra*.

23. In the king's absence, the country had been governed by a Council. On his return, Charles employed the services of the German Baron Heinrich von Görtz, previously deeply involved in political intrigues among Sweden's European enemies. Charles gave him a free hand in manipulating Sweden's economy to raise as much money as possible for further military exploits, which von Görtz did by diplomatic efforts abroad, oppressive domestic taxation and expropriation, printing emergency money, etc. He became one of the most hated figures in Swedish history, and his arrest and execution followed swiftly after Charles' death.

24. It should be remembered that Swedenborg's university degree did not qualify him for any particular post.

25. Swedenborg had written Latin poetry off and on since he was twelve. Both his 1710 *Festivus Applausus* celebrating Stenbock's victory and his 1715 *Festivus Applausus* celebrating Charles' dramatic return from Turkey are rhetorically sophisticated; and in 1715 he also composed *Camena Borea*, an allegorical poem describing Sweden's distress.

26. See Acton, *Letters and Memorials*, pp. 121f.

27. Swedenborg's own account of the discussions is found in his letter of 1734 to Jöran Andersson Nordberg, a priest and historian who served with Charles XII for many years during the European campaigns and was the king's confessor. See J. A. Nordberg, *Konung Carl XII:s Historia*, Stockholm 1740.

28. Polhem wrote to Charles that the College had need of "one... who understands mining ordinances.... therefore I submit to your Majesty's gracious decision whether this Swedberg—who has also qualified for a University Professor, may be advanced to the post of Assessor in the aforementioned College and therefore be kept in that field in which he is likely to be of greater service than at a University." See Acton, *op. cit.* pp. 126f.

29. See Tord Ångström, *Svenskt Flyg och dess män* (Stockholm 1939), pp. 44f.

IV   AFTER SWEDENBORG

1. See p. 25 *supra*. The position of Ulrika Eleonora (1688-1741) was somewhat unclear. In 1713, without the consent of her brother Charles XII (then imprisoned in Turkey), she assumed presidency of the governing Council. After Charles' death she was proclaimed regent in 1718 on the condition that she abolish absolute monarchy. Four days after her coronation, she ennobled 150 families, including Bishop Swedberg's. In 1720, she abdicated in favor of her husband, Frederic of Hessen. This may serve to suggest the chaotic political situation at the outset of Swedenborg's professional career.

2. In 1747, Swedenborg retired from his post on the College of Mines, just after his ability and his seniority had offered him the opportunity to be its head. By his own account, he had been preoccupied with thought about God from his childhood; and now his intense experiences of what he considered divine revelation convinced him that he was a chosen instrument of God. See Section 3 of the Bibliography.

3. In 1769, reports came from London on Swedenborg's situation and physical condition which were far from flattering. As an example, Professor Johan Hinric Lidin, a member of the Swedish Academy of Science, had visited with him, and pictured him as an old, sick man (he was then eighty-one) whose ideas bordered on madness. Such reports were received with malicious pleasure by many people in Sweden, especially in the State Church, which was consciously hostile to his religious ideas. See Cyriel Odhner, "Where are the Swedenborg Documents?," in *New Church Life*, Feb. 1925, pp. 89-91.

4. For the most part, his influence has been diffuse but very considerable. It

affected German Romantic philosophy (von Fichte, Carus), American transcendentalism (Emerson, Henry James Sr.), early socialism (de Saint-Simon, Fourier), and such novelists, poets, and artists as Goethe, Heine, Novalis, Balzac, Baudelaire, Coleridge, the Brownings, Strindberg, Blake, and Hiram Powers. See the article by Howard Davis Spoerl in *The Encyclopedia Americana* (1960 ed.), *s.v. Swedenborg.*

5. The Swedish king Gustavus III witnessed a balloon ascent at Lyon, France, as a guest of Louis XVI on June 4, 1784, and was most eager to provide his own people with a demonstration. The first flight in Sweden took place under his auspices from the "Observatory Hill" in Stockholm on September 17, 1784, the sole passenger being a cat.

6. "It is not impossible that Bauer had read or heard of the account of Swedenborg's Machine" (Gibbs-Smith, *Aviation*, p. 16). Gerhard Wissman, in his *Geschichte der Luftfahrt von Ikarus bis zur Gegenwart* (pp. 183-4). points to Bauer's design as closely resembling Swedenborg's in basic principles, but does not speculate about any direct influence. Since the Swedenborg papers were still undiscovered in the Library at Linköping, the only possible link would have been a copy of the *Daedalus* version, which is conceivable but unlikely. Clive Hart devotes considerable attention to Bauer in Chapter 9 of his *Prehistory of Flight.*

7. Cayley's three epoch-making articles, under the general title "On Aerial Navigation," were published in November 1790, February 1810, and March 1810.

8. In a letter to his friend and benefactor Viscount Mahon (later the fourth Lord of Stanhope), Cayley writes in response to the Viscount's commendation that no one has gone further than himself in experiments, but that he wants to refer to hundreds of precursors through the ages from Daedalus. See Gibbs-Smith, *Sir George Cayley's Aeronautics 1796-1855* (London 1962), pp. 60f.

9. Two biographies of Cayley of particular interest are *The Life of a Genius* by Gerald Fairlie and Elizabeth Cayley (London 1965) and *Sir George Cayley, the Inventor of the Aeroplane* by J. Lawrence Pritchard (London 1961). Neither finds evidence that Cayley was aware of Swedenborg or his design.

10. This was an important event in the history of aeronautics. The members were men of established scientific reputation, devoting themselves exclusively to the problems of flight. See C. C. M. Brown, *The Conquest of the Air* (London, 1927), pp. 56-60.

11. The first meeting of the Royal Aeronautical Society was held on June 27, 1866. Wenham's lecture was entitled "Aerial Locomotion and the Laws by which Heavy Bodies Impelled through the Air are Sustained," and presented the results of significant tests he had performed in 1858 and 1859. These results were widely used by later inventors.

12. Lilienthal was of Swedish origin. One of his ancestors had been an officer in the armies of King Gustavus Adolphus II during the Thirty Years' War in Europe. He did not return to mainland Sweden but settled in Pomerania, which later became part of Germany (see p. 29 *supra*).

13. In his highly respected work *Aerodynamics* (Ithaca, N.Y. 1954), Theodore von Karman writes, "Lilienthal strongly emphasized the importance of curved

wing surfaces. He made many other interesting aerodynamic observations: he found, for example, that natural winds are more favourable for soaring flight than a perfectly uniform airflow. This favourable effect can be achieved by utilizing the upward components which often exist in the natural wind. Lilienthal found, however, that sometimes the lift in natural wind, even in the absence of upward components, may be superior to that measured in a uniform airstream. Only in recent times was it recognized that this effect is due to a cross-velocity gradient, which generally prevails in the natural wind, at least in lower layers of the atmosphere."

14. For obvious historical reasons, the Americans were late arrivals on the scene. One significant figure was Sir Hiram Maxim, an American businessman who moved to England, where he was eventually knighted. While he was not an independent inventor, he gave generous financial support to a number of experiments and took a lively interest in the history of flight.

15. Chanute, a French born American, was originally a railroad engineer, instrumental in the construction of the elevated transit systems of New York and Chicago. From the age of 51, he devoted all his time to flying, becoming widely known not only for his own experiments, but for his activity as consultant to aviation pioneers in many countries. His correspondence fills twenty-two volumes in the Library of Congress in Washington. Gibbs-Smith, in his 1953 *History of Flying*, writes, "Chanute's book 'Progress in Flying' became one of the Bibles of aeronautics. It was the first recorded analysis of aviation experiments ever to appear, and a book which is symbolic of Chanute's chief function in history. For he was the great collector and disseminator of accurate information about aviation as well as being a pioneer in his own right."

16. See pp. 47ff. *supra*.

17. A second effort on December 8, 1903, was even less successful, and the United States government withdrew its support. Only nine days after this second failure, the Wright brothers made their historic flight. A year after Langley's death, a special Langley Association was formed, and through its efforts a model of "The Aerodrome" now stands in the Smithsonian's Air and Space Museum, not far from the model of Swedenborg's craft.

V EXPERTS' EVALUATIONS

1. See p. 22f. *supra*.

2. Carl von Klinckowström, "Emanuel Swedenborg und das Flugproblem," in *Geschichtsblätter füer Technik, Industrie und Gewerbe* Nr. 7-9 und 10-12, Jahrgang 1916 (Berlin). von Klinckowström is primarily known as a historian of cultural and technical subjects and as an editor of technical journals. He had a position on the German General Staff after having been wounded in World War I, and it was during his work there that he wrote his article on Swedenborg. After the war he lived in Munich where he worked as a teacher, lecturer, publisher, and librarian.

3. Polhem qualified his objections with the remark, "However, it would be well to take advantage of a wind, if the same were constant and invariable."

Swedenborg himself had made the same point; the first of his "four requisites" was "a strong wind," and he added that "in calm weather it would be better to keep quietly and humbly by the ground." See p. 22 *supra*.

4. *op. cit.*, p. 227.

5. See pp. 42f. *supra*.

6. See pp. 56f. *supra*.

7. See *A Memorandum of the Relationship of the Proposed Swedenborg Flying Machine to Certain Early Proposals of the Flying Machine of the Nineteenth Century*, in the correspondence file of the library of the Smithsonian Air and Space Museum.

8. *Aileron*: a movable part of an airplane wing or a movable airfoil external to the wing at the trailing edge for imparting a rolling motion and thus providing lateral control (*Webster's Ninth New Intercollegiate Dictionary*, 1983, *s.v.*).

9. M. A. Goupil published his theories in the book, *La Locomotion Aérienne* in 1884.

10. On Mattulath, see Gibbs-Smith, *Aviation, An Historical Survey* (1970), Vol. II.

11. It should be remembered that old Swedish presents difficulties to the modern reader, and that Swedenborg's handwriting can be difficult to decipher.

12. This problem was also noted by Gustaf Genzlinger, who built the model for the Smithsonian. After consulting with experts at the museum, he reached the same conclusion as did Holmer. See p. 53f. *supra*.

13. See p. 34 *supra*.

VI DOCUMENTATION

1. The reason for the archive's being at Linköping is that Swedenborg's brother-in-law Eric Benzelius was bishop there from 1731 to 1742. He brought most of his records with him from Upsala, including his letters from Swedenborg.

2. See Greta Ekelöf, "The History of Swedenborg's manuscripts preserved in the library of the Royal Swedish Academy of Science and Other Swedish Libraries," *Transactions of the International Swedenborg Congress* (London 1910), pp. 337-52. See also Alfred Stroh and Greta Ekelöf, "En förkortad kronologisk förteckning över Emanuel Swedenborgs arbeten," *Kgl. Vetenskapsakademins årsbok* (1910); Sigrid Cyriel Odhner, "Where are the Swedenborg Documents?", *New Church Life* (February 1925); Arne Holmberg, "Swedenborg Manuscripts," *New Church Magazine*, (July-September 1938); "The Swedenborg Documents," *New Church Life* (January 1927).

3. See James J. G. Wilkinson, *Emanuel Swedenborg—A Biography* (London 1849).

4. See Immanuel Tafel, *Sammlung von Urkunden betreffend das Leben und den Character Emanuel Swedenborgs* (Tübingen 1839).

5. Rudolf L. Tafel, *Documents concerning the Life and Character of Emanuel Swedenborg* (London 1875). The manuscript on the flying machine is included as paragraph 6 in Document 311, pp. 876-9. Tafel was a Professor at Washington University in St. Louis, Missouri.

6. The Academy was founded on June 2, 1739. On December 10, 1740, Swedenborg was unanimously elected as its fifty-second member. Had he not been abroad at the time of its founding, he would doubtless have been one of its charter members, especially since he had advocated such a society more than twenty years before (see p. 60 *supra*). As it was, he was elected immediately after his return to Sweden in October 1740 after a four and a half year journey abroad.

7. The committee was proposed by Gustaf Retzius, Professor of Medicine. It consisted of Retzius (Medicine-Anatomy), Ch. Lovén (Medicine-Anatomy), A.G. Nathorst (Geology), S.E. Henschen (Medicine-Brain), and S. Arrhenius (Physics), with A. Stroh as Secretary. In a letter to Stroh published in the April, 1903 issue of *The New Philosophy*, Retzius describes the conception of the committee.

8. The three volumes were published in 1907, 1908, and 1909 under the collective title, *Opera Quaedam Aut Inedita Aut Obsoleta De Rebus Naturalibus*, Ed. Alfred H. Stroh; Preface by Gustaf Retzius.

9. During this period, Stroh scrupulously reported back to his principals, the Swedenborg Scientific Association in Philadelphia. His more significant letters, describing the progress in the search for documents in Sweden, were periodically reproduced in the Association's journal, *The New Philosophy*, mainly between 1902 and 1911.

10. In *The History of Flying*. Gibbs-Smith gives the following list of events in Europe between the Wright brothers' first flight in 1903 and the Swedenborg Congress in 1910.

1903	(17 December) First powered, controlled and sustained aeoplane flight, the Wright brothers
1903	Europe receives first detail of Wright and Chanute gliders and restarts experiments
1905	Wrights fly 24⅓ miles nonstop in 38 minutes 3 seconds
1905	First "box-kite" float glider flies (Archdeacon-Voisin)
1906	Wrights' powered flights of 1903-5 first admitted and made known to the public
1906	First powered "box-kite" aeroplane flies—also the first official aeroplane flight in Europe (Santos-Dumont)
1906	First Godon-Bennett balloon race
1907	First full-size monoplane flies (Bleriot)
1907	Delagrange-Voisin biplane flies
1907	First man-carrying helicopter flights (Breguet brothers and Cornu)
1907	Lanchester publishes his Aerodynamics
1907-8	First flying bomb proposed (Lorin)
1908	The Wrights' first flights in public: Wilbur in Europe, Orville in USA
1908	First aeroplane fatality (Lt Selfridge, when passenger with Orville Wright)
1908	First official aeroplane flight in England (Cody)
1909	Henri Farman produces his first aeroplane
1909	First aviation meeting held (at Rheims)
1909	First Channel crossing by aeroplane (Bleriot)
1910	First international air race, London to Manchester

1910   Junkers obtains patent for an "all-wing" airplane
1910   First woman qualifies as aeroplane pilot (Baronesse de Laroche)
1910   First air passenger service with a Zeppelin airship (Deutsche Luft-schiffahrt A.G.)

VII "THE MACHINE" IN AVIATION LITERATURE

1. Hugo Lj. Odhner and Carl Th. Odhner, *Suggestions for a Flying Machine* (Philadelphia 1910). See pp. 17ff. *supra.*

2. Alfred Acton, "Suggestions for a Flying Machine," in *The Scientific American* (October 28, 1911) Supplement No. 1869, pp. 286f. Acton was a leading scholar in the Swedenborgian community outside Philadelphia and therefore a colleague of the Odhners.

3. T. O'B. Hubbard and C.C. Turner, *The Boys' Book of Aeroplanes* (London 1912), contains a picture and description. Hubbard was Secretary General of the Royal Aeronautical Society, and in this publication makes reference to the article in the Society's *Journal.* See p. 32 *supra.*

4. Carl von Klinckowström, "Emanuel Swedenborg und das Flugproblem," in *Geschichtsblätter für Technik, Industrie u. Gewerbe*, Nr. 7-9 u. 10-12, Jahrgang 1916. See p. 32 *supra.*

5. See p. 32 *supra.*

6. Tord Ångström, "Om flygning och Luftfahrt," Bonniers: *Orientering i aktuella ämnen* (Stockholm 1934), pp. 10-14.

7. _____, *Machine att flyga i Wädret*, Publ. "Facsimilia" (Stockholm 1960), Preface by S. Rosen and comments by O. Hjern; translation reprinted from Acton, "Mechanical Inventions," (1939), pp. 20-26.

8. Jules Duhem, *Histoire des idées aéronautiques avant Montgolfier* (Paris 1943), pp. 12, 245. Duhem notes that many inventors also had the idea of a surface to serve as a parachute in case of emergency.

9. In his *Musée Aéronautique* (Paris 1944), pp. 160, 163, a pictorial presentation is accompanied by extensive comments. Duhem proposes influences from the Marquis de Worcest's *A Century of Inventions* and Hooke's *Philosophical Collections.*

10. C.L.M. Brown, *The Conquest of the Air* (London 1927), p. 38. Brown refers to Swedenborg as a "restless, erratic genius" who wandered in strange fields of thoughts, who invented an aeroplane which was "cumbersome" and "impracticable." Swedenborg himself would probably have endorsed Brown's statement that it was "...interesting only so far as it shows that men were developing an increased capacity to grasp the difficulties of the problems before them... ."

11. Alfred Acton, *The Mechanical Inventions of Emanuel Swedenborg* (Philadelphia 1939). He mentions (in part following Swedenborg) Wilkins, Kircher, Lowthrop, and Bacon as sources of Swedenborg's inspiration.

12. See p. 74 *supra.*

13. See, e.g., *Newsweek* 11:25, *Literary Digest* 125:21, *Science Monthly* 125:21, *Nineteenth Century* 123: 328-33, *Time* 31:40-1, *Science* sup. 10, all in 1938.

14. John Goldstrom, "He Foresaw the Aircraft," in *The Swedish-American Monthly* (New York 1938) [*Issue No. and Month?*]

15. *ibid.*

16. See pp. 47ff. *supra.*

17. While Goldstrom expresses his intent to return to the subject of "the incredible Swede," he never, to this author's knowledge, published more on the topic.

18. See Henry Söderberg, "'Rediscovery' of Swedenborg's Machine to Fly in the Air." *Logos,* Newsletter of the Swedenborg Foundation, Fall 1986 (New York).

19. In *A History of Flying* (1953), he takes the view that the chief interest in the description, "which talks 'vaguely' about movements and control, lies in the 'hood' and the device for stabilizing the machine" (pp. 57f.). In *A Historical Survey* (1970), he seems more skeptical, stating that "coming from a man of such eminence, this idea is interesting, but except for the pendulum stability—without merit historically" (pp. 14ff.). He may have meant "practically" rather than "historically," which would better suit the facts.

20. See pp. 28, and 72 *supra.*

21. See p. 42 *supra.*

22. See Chapter Eight *infra,* "Models of the Machine."

23. See Gustaf Genzlinger, "A Model of Swedenborg's Flying Machine," *The New Philosophy* (April 1982), pp. 33-39. The article gives the background of the model, and also provides general information about the airplane itself.

24. Gerhard Wissman, *Geschichte der Luftfahrt* (Muenchen 1960). The Swedenborg airplane is discussed on pp. 150ff.

25. See p. 42 *supra.*

26. Ernst Benz, *Emanuel Swedenborg* (Zürich 1969). The airplane is discussed on pp. 30 and 61.

27. See p. 3 *supra.*

28. Leslie H. Hayward, *The History of Air Cushion Vehicles* (London 1963). The material cited above is found on pp. 3f. Hayward's earlier works on helicopters earned him the Bronze Medal of the Swedish Society of Aeronautics in 1962.

29. *Automobile Quarterly* (Winter 1967), p. 285. "Although Swedenborg seems to have recognized that his machine could not really support itself (adequate mechanical power simply was not available and the man-powered oars would beat the air futilely) he must nevertheless be acknowledged as man's first attempt to utilize the cushioning effects of air."

VIII MODELS OF "THE MACHINE"

1. See p. 65 *supra.*

2. The efforts of Immanuel and Rudolf Tafel and of Alfred Stroh are particularly noteworthy. See pp. 37 and 65 *supra.*

3. A Swedenborg Society was founded in London in 1862, and now has excellent archival resources. The Swedenborg Scientific Association in Bryn Athyn, Pennsylvania (est. 1898) is a primary publisher of Swedenborg's scientific and

NOTES

philosophical works; and the Swedenborgian Archives at Bryn Athyn are a particularly rich source of information.

4. See pp. 47ff. *supra*. Reports that Burt's source was Swedenborg's *Principia* cannot be taken seriously, since the *Principia* makes no mention of the airplane. The error is probably due to the fact that among Swedenborgians, the *Principia* is the best known of his scientific works and would come first to mind. *Cf.* "It's probably from Shakespeare or the Bible."

5. See the letter of June 5, 1947, from Marguerite Block of Columbia University to Thorsten Althin, Curator of the Technical Museum in Stockholm (Archives, Technical Museum, Stockholm).

6. Letter undated, probably 1898, from Mrs. Grace Burt Böricke to Dorothy Burnham (Archives, Glenview Academy).

7. The cockpit has been a subject of debate. See pp. 54f. *supra*.

8. *Offene Tore* (Zürich, Swedenborg Verlag), 1960. (The calendar was published by Scandinavian Airlines System, SAS.)

9. The story, by Alexander McQueen, dealt mainly with the so-called "Venetian-Blind Airplane" produced by Horatio Philips in 1892. Philips was a well-known English inventor of the nineties; he laid the modern foundations of airfoil design. His invention was therefore related to Swedenborg's in both design and construction, giving McQueen the occasion to refer to the machine and to the Glenview experiment. He also refers to the model in the Smithsonian (Archives, Library, Smithsonian Air and Space Museum, Washington, D.C. *s.v.* Swedenborg).

10. See pp. 37f. *supra*.

11. *Transactions of the International Swedenborg Congress, 1910* (London 1910), p. 5.

12. *ibid.*

13. Swedenborg House in London (the headquarters of the Swedenborg Society) informs me that there are no known pictures or records of the model, which is assumed to have been small enough to be readily portable.

14. See pp. 32 *supra*.

15. Late in 1986, the author queried several German aviation institutions about the Amman model. Its whereabouts could not be traced. The most likely location for it is the Deutsches Museum von Meisterwerken der Naturwissenschaft und Technik in Munich; but a letter from the Museum's Administration (dated January 17, 1987) states that they have been unable to find the model and have no records of Herr Amman. The model and any records may not have survived two world wars.

16. See p. 32 *supra*.

17. Tord Ångström, in *Svensk flyg och dess män* [*Swedish Flying and its Men*] (Stockholm, Royal Swedish Aeroclub, 1939), pp. 42-46.

18. Alfred Acton, *op. cit.*.

19. Lennart Alfeldt, review of Tord Ångström's *Machine att flyga i Waedret*, in *The New Philosophy* (Bryn Athyn 1961).

20. Tord Ångström, *Machine att flyga i Wädret* (Stockholm, Facsimilia, 1960); Preface by S. Rosen and comments by O. Hjern.

21. Gustav Genzlinger, "A Model of Swedenborg's Flying Machine," *The New Philosophy*, April-June 1962, pp. 33-40.

22. From the Archives of the Library of the Smithsonian Air and Space Museum.

23. Genzlinger, *op. cit.*.

24. Gunnar Jarring (1907-), internationally known Swedish diplomat, expert in Oriental languages, Ambassador to the United Nations 1956-58, Washington 1958-64, and Moscow 1964-73.

APPENDIX

1. Most of the information about his meetings with learned men during this first trip abroad comes from his letters to Eric Benzelius. See Alfred Acton, *Letters and Memorials* (Bryn Athyn, 1948), pp. 12-55. This Appendix is a somewhat condensed and rearranged version of that information.

2. See pp. 8 and 11 *supra*.

3. Inspired perhaps by his contacts with Flamsteed and his interest in Halley, he devised a system for finding longitude at sea by observing the moon. Though he kept trying all his life, he never won acceptance for his proposal, to his considerable disappointment. During his stay in Paris, he tried hard both through Abbé Bignon and through Varignon to have his findings on longitude submitted to the Royal French Academy. He was evidently annoyed when he did not succeed.

4. There was, under Charles II, a particular interest in solving the problem of finding longitude at sea, England being commercially dependent on her merchant fleets.

5. See pp. 20f. *supra*.

6. *Hist. Acad. Royal des Sciences.* 1743, p. 189.

7. See pp. 5 and 65 *supra*.

BIBLIOGRAPHY

Thanks partly to his own extensive self-documentation and partly to the research of his followers, Emanuel Swedenborg is unusually well covered by bibliographies and special studies. Comprehensive articles can be found *s.v.* "Swedenborg" in all reputable encyclopedias.

For present purposes, the following bibliography is presented in three sections:

1. Literature which refers directly to Swedenborg's machine and which adds to our information about or understanding of it.

2. Selected literature on the history of aviation directly relevant to Swedenborg's machine. It should be noted that this has become an extensive body of literature. In the Library of the Smithsonian National Air and Space Museum, for example, there are some 700 titles listed under the heading, "History of Aviation." My listings under this heading should be regarded as highly selective.

3. Selected literature about Emanuel Swedenborg, his life, work and character, as background to the subject of this book. The Library of Congress lists about 100 titles under this heading. I have not included Swedenborg's own works.

The works are listed in chronological rather than alphabetical order so as to demonstrate the development of literature in relation to Swedenborg's airplane.

1. LITERATURE REFERRING DIRECTLY TO SWEDENBORG'S AIRPLANE

Swedenborg, Emanuel. *Manuscript: Machina Volatilis et Daedaleae*. With drawing. Probably produced 1714. In Linköpings Stiftsbibliotek, Codex 14a. (English translation, the Diocesan Library of Linköping).
————. "Utkast till en Machine att flyga i wädret," *Daedalus Hyperboreus* 4, (Oct.-Dec. 1716) pp. 80-83.
Tafel, Rudolf L. *Documents concerning the Life and Character of Emanuel Swedenborg*. London: Swedenborg Society, 1875-77. 3 Vol's.
Swedenborg, Emanuel. *Opera Quaedam aut Inedita aut Obsoleta de Rebus*

Naturalibus I. Ed. Alfred H. Stroh. Stockholm: 1907. Includes letters about a flying machine, pp. 224-29.

―――――. "Swedenborg's Flying Machine," *New Church Life*, (October 1909) pp. 582-591. Translated with commentary by Carl Th. Odhner.

―――――. *Suggestions for a Flying Machine*. Philadelphia: 1910. Translated by Hugo Lj. Odhner and Carl Th. Odhner.

―――――. "Suggestions for a Flying Machine." *The Journal of the Royal Aeronautical Society*, July 1910, pp. 118-22. Foreword by T. O'B. Hubbard, Editor. Reprint of the preceding item.

Dunér, Nils. Facsimiles of *Daedalus Hyperboreus* vols. I-IV. *Kongl. Vetenskapssocietetens i Upsala Tvåhundraårsminne*. Stockholm, Generalstabens Litografiska, 1910.

Anon. "Swedenborg as inventor of Aeroplane," *Dagens Nyheter*, July 7, 1910.

Transactions of the International Swedenborg Congress. London: Swedenborg Society, 1910, pp. 5, 45f. and Fig. 1.

Svenska Teknologföreningens Minneskrift. Polheim, Christoffer. A Biography. Stockholm: 1911.

Swedenborg, Emanuel. "Suggestions for a Flying Machine," *Scientific American* (October 28, 1911), Supplement No. 1869, pp. 286f. With introduction by Alfred Acton.

Hubbard, T. O'B. and Turner, C. C. *The Boys' Book on Aeroplanes*. London: 1912. p. 105. Description and picture.

Klinckowström, Carl von. "Emanuel Swedenborg und das Flugproblem," *Geschiahtsblätter für Technik, Industrie, u. Gewerbe* Nr. 7-9 u. 10-12 (Jahrgang 1916). Berlin.

Vivian, E. C. and Marsh, W. Lockwood. *A History of Aeronautics* (London: Harcourt, Brace & Co. 1921).

Ångström, Tord. "Early trials to solve the problem of flying," *Uppfinnigarnas bok* ("The book of Inventions"). Stockholm: P. A. Norstedt, 1925, pp. 1129-1135. 2nd ed. 1931, pp. 1073-1247

Goldstrom, John. *A Narrative Story of Aviation*. New York: Macmillan, 1930.

Brown, Cecil L. M. *The Conquest of the Air*. London: Oxford University Press, 1927, p. 38.

Klinckowström, Carl von. "Schwingenflieger," *Münchener Neueste Nachrichten*. January 19, 1930.

Ångström, Tord. "Emanuel Swedenborgs flygplansprojekt," *Daedalus, Yearbook of the Technical Museum, Stockholm*. 1932.

―――――. "Om flygning och luftfart," *Bonniers: Orientering i aktuella ämnen*. Stockholm: 1934. pp. 10-14.

Lüdecke, Heinz. *Vom Zaubervogel zu Zeppelin*. Berlin: Kurt Wolff Verl., 1936, pp. 161-66, 183.

Goldström, John. "He foresaw the aircraft," *The American-Swedish Monthly*, February 1938. New York.

Wales, Margaret. "Emanuel Swedenborg, 18th Century Pioneer," *Popular Aviation*. Chicago: 1938.

Dufty, J. G. *Swedenborg, the Scientist*. London: The Swedenborg Society, 1938.

Anon. Photograph of the model in the Stockholm Technical Museum. *Teknisk Tidskrift*, May, 1938.

Acton, Alfred. *The Mechanical Inventions of Emanuel Swedenborg*. Philadelphia, Swedenborg Scientific Association, 1939.

Ångström, Tord. In *Svenskt flyg och dess män*, for the benefit of the Royal Swedish Aeroclub. Stockholm: Bokförlaget Mimer, 1939. pp. 42-45.

Duhem, Jules. *Histoire des idées Aeronautiques avant Montgolfier*. Paris: Fernand Sorlot, 1943.

————. *Musée aéronautique avant Montgolfier*. Paris: Fernand Sorlot, 1943. p. 108.

Block, Marguerite. Letter concerning Jesse Burt's glider (1898), to Thorsten Althin, Curator, Technical Museum, Stockholm, dated June 5, 1947. In the Archives of the Museum.

Acton, Alfred, ed. *Letters and Memorials*. Bryn Athyn, PA: The Swedenborg Scientific Association, 1948. pp. 55ff, 124, 135.

Gibbs-Smith, Charles H. *A History of Flying*. London: B. T. Batsford Ltd., 1953. pp. 57f.

Supf, Peter. *Die Eroberung des Luftreichs*. Stuttgart: Konradin Verlag, 1953. pp. 146f., 154.

Wissman, Gerhard. *Geschichte der Luftfahrt*. Berlin: VEB Verlag Technik, 1960. pp. 150-52.

Ångström, Tord. *Machine att flyga i Wädret*. Stockholm: Bokförlaget Facsimilia, 1960. Preface by S. Rosén and comments by O. Hjern.

The New Church Messenger, May 1960. Reproduction of and comments about the model of the Swedenborg machine in the Stockholm Technical Museum.

Alfeldt, Lennart. Review of Ångström's *Machine att flyga i Wädret* (see above). *The New Philosophy* (Bryn Athyn: Swedenborg Scientific Association, 1961).

The Story of Jesse Burt's Model. "Venetian-Blind Aeroplane." Cover of unidentified calendar, no date. In the archives of the Glenview Academy, Ill.

Peebles, Waldo. "Pioneer of Flight," translated from German. *The New Church Messenger*, May 1960.

National Broadcasting Corporation. Station WRC radio interview (transcript) with the Swedish Ambassador to Washington, Gunnar Jarring, on January 29, 1962, in connection with the presentation of a model of the machine to the Smithsonian Air and Space Museum. (Smithsonian Archives, Washington, D.C.).

Press Release, January 29, 1962. Dealing with the presentation of the model to the Smithsonian Air and Space Museum. (Smithsonian Archives, Washington, D.C.).

Douglas, Earl L. "Religious Off-Beat." Article about Emanuel Swedenborg in connection with the presentation of the model to the Smithsonian Air and Space Museum. In *Christian Herald*, August 1961.

Views on the model of the machine Letter of April 20, 1962, from the Curator of the Smithsonian Air and Space Museum, Paul E. Garber, to Gustaf Genzlinger, the builder of the model, with viewpoints and suggestions. (Smithsonian Archives, Washington, D. C.).

BIBLIOGRAPHY

Genzlinger, Gustaf. "A Model of Swedenborg's Flying Machine." *The New Philosophy*, April-June 1962 (Philadelphia).
Hayward, Leslie. *The History of Cushion Vehicles*. London: 1963, pp. 3f.
Automobile Quarterly, Vol. 5 No. 3, Winter 1967. Article on the cushioning effect of air in relation to Swedenborg's machine.
Benz, Ernst. *Emanuel Swedenborg*. Zürich: Swedenborg Verlag, 1969, pp. 30, 61.
Flygets Årsbok (Allhem). Swedish yearbook on aviation. References to Swedenborg's machine: 1945, p. 26; 1969, p. 263; 1976, p. 13; 1985, p. 155.
Gibbs-Smith, Charles H. *Aviation: An Historical Survey from its Origins to the End of World War II*. London: H. M. S. O. 1970, pp. 14, 16.
Seelman, Wolf Dietr. *Illustrierte Geschichte der Fliegerei*. Berlin: Verl. M. Pawlak Herrsching, 1970.
Atterlid, Tore. "Emanuel Swedenborgs Flygplansprojekt," *VIPS, SAAB Personnel Magazine* July 1977 (Linköping).
_____. "Kring Swedenborgs Flygplansprojekt," *Östgöta Correpondenten*, July 13, 1977 (Linkoeping).
Hart, Clive. "Swedenborg's Flying Saucer," *Journal of the Royal Aeronautical Society*, September 1980 (London).
Nilsson, C.-G. "Hur flyger fåglarna?" *Fåglar i Norrbotten*, No 4, 1984 (Stockholm).
Hart, Clive. *The Prehistory of Flight*. London: University of California Press, 1985.
Falk, Bertil. "Emanuel Swedenborgs Flygmaskin från 1716," *Officersförbundsbladet*, No. 9, 1986 (Stockholm).
Söderberg, Henry. "Rediscovery of Swedenborg's Machine to Fly in the Air," *Logos*, Swedenborg Foundation Newsletter, Fall 1986 (New York). Interview.
_____. "Swedenborg—Sweden's first aeronautical engineer?." The Saab-Scania Griffin, Linköping, 1987.

2. SELECTED LITERATURE ABOUT THE HISTORY OF AVIATION WITH SPECIAL BEARING ON SWEDENBORG'S AEROPLANE

Bacon, Roger. *De Mirabili Potestate Artis et Naturae* (Written about 1250). Paris: Lutetia Parisiorum, 1542.
Vinci, Leonardo da. *Codice sur Volo Degli Uccelli* (1504). Transcription and Translation by G. Piumati and C. Ravaisson-Mollien (Parigi: 1893).
_____. *Les Manuscrits de Léonardo da Vinci... facsimilés....* 6 Vols. Paris: 1881-1891. With transcriptions and French translations. ed. C. Ravaisson-Mollien.
Le Journal des Scavans Articles about Danti and Besnier. Paris: December 1678.
Borelli, Giovanni. *De Motu Animalium*, 2 Vols. Rome: 1680–81.
Rousseau, Jean Jacques. *Le Nouveau Dédale* (1742). Ed. Pierre-Paul Plan, *Mercure de France* 87, 4 (1 October 1910), 577-97.
Meerwein, Carl Friedrich. *Kunst zu Fliegen nach Art der Vögel*. Frankfurt/Basel: J. J. Thourneysen Fils, 1784. French version published as *L'Art de voler à la Manière des Oiseaux* (Basel: 1784); Portuguese version published as *A Arte de Voar a' Maniera dos Passaros* (Lisboa: 1812).

Cayley, Sir George. "On Aerial Navigation" (Mechanical Flights). Articles in *Nicholson's Journal of Philosophy*. London: 1809-10. Vols. XXIV and XXV reprinted in *The Aeronautical Classics Series*, No. 1, 1910.

Wenham, F. H. "Aerial Locomotion." First Annual Report of the Royal Aeronautical Society. London: 1866. Reprinted in *The Aeronautical Classics Series*, No. 2, 1910.

Lilienthal, Otto. *Der Vogelflug als Grundlage der Flieger-Kunst* ("Birdflight as the Basis of Aviation"). Berlin: 1889. London: Longmans, Green & Co., 1911.

Chanute, Octave. *Progress in Flying Machines*. New York: 1894. Facsimile Edition, Long Beach, CA: Lorens and Herweg, 1976.

————. Papers, 1850-1910. Library of Congress, Manuscript Division. 21,000 items.

Alexander, John. *The Conquest of the Air*, with Preface by Hiram Maxim. New York, S. W. Partridge & Co., 1902.

Wilhelm, Balthasar, S. J. *Die Anfänge der Luftfahrt, Lana-Gusmão. Zur Erinnerung an den 200. Gedenktag des ersten Ballonaufstieges, 8. Aug. 1709*. Hamm i. Westphalen: Breer & Thiemann, 1909.

Klinckowström, Carl von. "Luftfahrten in der Literatur," *Zeitschrift für Bücherfreunde*, no. 3.2. Munich: 1911. pp. 250-264.

Brown, C. L. M. *The Conquest of the Air*. London: 1927.

Walker, Ernest E. *Aviation, The Story of the Conquest of the Air*. Chicago: Aeronautics Education Foundation, 1927.

Black, Archibald. *The Story of Flying*. New York: McGraw Hill, 1940.

Gibbs-Smith, Charles H. *A Directory and Nomenclature of the First Aeroplanes 1809 to 1909* . London: H. M. S. O., 1953.

De Leeuw, Hendrik. *The Conquest of the Air: The History and Future of Aviation*. New York: Vantage, 1959.

Pritchard, J. Lawrence. *Sir George Cayley—The Inventor of the Aeroplane*. London: Max Parrish, 1961.

Gibbs-Smith, Charles H. *Sir George Cayley's Aeronautics 1796-1855*. London: H. M. S. O., 1961.

Hart, Clive. *Kites: An Historical Survey*. London: 1962. Revised Ed. 1982.

Fairlie, Gerald and Cayley, Elizabeth. *The Life of a Genius*. London: Hodder and Staughton, 1965.

Gibbs-Smith, Charles H. *The World's First Aeroplane Flights*. London: H. M. S. O., 1965.

Rundberg, A. *100 år i luften*. Tiden, Stockholm: 1965.

Stewart, Oliver. *The Conquest of the Air*. New York: H. M. S. O., 1972.

Taylor, John W. R. *History of Aviation*. New York: Crown, 1972.

Hart, Clive. *The Dream of Flight: Aeronautics from Classical Times to the Renaissance*. London: Faber & Faber, 1972.

Gibbs-Smith, Charles H. *Flight through the Ages*. New York: Crowell, 1974.

3. SELECTED LITERATURE ABOUT EMANUEL SWEDENBORG AS BACKGROUND TO THE CONTENTS OF THE BOOK

Lidén, Joh. Hinr. *Repertorium Benzelanium*. Inventory of manuscripts in the Linköpings Stiftsbibliotek. Stockholm: Tryckrhos Anders J. Nordström, 1791.

Schröder, J. H. *Histoire de la Societé Royale de Sciences d'Upsala.* Upsala: Leffler & Sebell, 1846.

Wilkinson, James J. G. *Emanuel Swedenborg: A Biography.* London: W. Newbury, 1849.

Tafel, J. F. Immanuel. *Sammlung von Urkunden betreffend das Leben und den Character Emanuel Swedenborgs* Four vol's. Tübingen: Tübingen Verlagsexpedition, 1839-45.

White, William. *Emanuel Swedenborg, His Life and Writings.* London: Simpkin, Marshall & Co., 1868.

Tafel, Rudolf L. Documents concerning the Life and Character of Emanuel Swedenborg. Three vol's. London: Swedenborg Society, 1875-77.

Clarke, F. W. *Swedenborg's Contribution to Science.* Address on the Atomic Theory before the Manchester Philosophical Society (1903).

Emerson, Ralph Waldo. *Representative Men: Seven Lectures.* London: Macmillan & Co., 1906.

Hyde, James J. G. *A Bibliography of the Works of Emanuel Swedenborg.* London: Swedenborg Society, 1906.

Stroh, Alfred H.. Ed. *Opera Quaedam aut Inedita aut Obsoleta de Rebus Naturalibus.* Stockholm: 1907.

Arrhenius, Svante. *Emanuel Swedenborg as a Cosmologist.* Stockholm; Aftonbladets Tryckeri, 1908.

Stroh, Alfred H. *Grunddragen av Swedenborgs lif.* Stockholm: Almqvist & Wiksell, 1909.

_____. "Outlines of Swedenborg's Early Life," in *New Church Life*, December 1909 (Bryn Athyn. PA).

_____. An Abridged Chronological List of the Works of Emanuel Swedenborg. Upsala and Stockholm: Almqvist & Wiksell, 1910.

Ekelöf, Greta. "The History of Swedenborg's Manuscript," in *Transactions of the International Swedenborg Congress 1910.* London: Swedenborg Society, 1910, pp. 337-354.

Lamm, Martin. *Swedenborg.* Stockholm: Gebers/Hammarstroem & Åberg, 1915, 1987.

Eby, Samuel C. *The Story of the Swedenborg Manuscripts.* New York: New Church Press, 1926.

Dingle, Herbert. *Swedenborg as a Physical Scientist.* Address at Queens Hall, January 28, 1938. London: Swedenborg Society, 1938.

Eby, Samuel C. *Alfred Stroh's Service to Swedenborg.* Address to General Conference (Swedenborgian). London: Swedenborg Society, 1931.

Bengtsson, Frans G. *Karl XII:s Levnad.* Malmö: P. A. Norstedt, 1935, pp. 308-310.

Holmberg, Arne. (Chief Librarian, Royal Swedish Academy of Science). "Swedenborg Manuscripts." Address at a meeting at Swedenborg House, London, January 28, 1938. *The New Church Magazine*, July-Sept. 1938.

Hildebrand, K. G. "Swedenborg och Karl XII," in *Svenska Dagbladet*, 30 November 1947.

Benz, Ernst. *Swedenborg in Deutschland.* Frankfurt am Main: V. Klostermann, 1947.

————. *Emanuel Swedenborg*. Munich: H. Rinn, 1948. Second Ed. Zürich: Swedenborg Verlag, 1969.

Toksvig, Signe. *Emanuel Swedenborg, Scientist and Mystic*. New Haven, Yale University Press, 1948.

Rydberg, S. *Svenska Studieresor till England under frihetstiden*. Upsala: Almqvist & Wiksell, 1951. With summary in English.

Andersson, I. *A History of Sweden*. London: Weidenfeld & Nicolson, 1956.

Academy Collection of Swedenborg Documents, Vol's. I-X, containing every known document by or concerning Swedenborg, including all available transcriptions and translations thereof. (Bryn Athyn, PA: 1962).

Hatton, Ragnhild M. *Charles XII of Sweden*. London: Historical Association, 1974.

Lindroth, Sten. "Emanuel Swedenborg," in *Swedish Men of Science*. Stockholm: Almqvist & Wiksell, 1952.

Wahlund, Per Erik. *Drömboken*. Stockholm: Wahlström & Widstrand. 1964.

Jonsson, Inge. *Emanuel Swedenborg*. New York: Twayne, 1971.

————, and Hjern, Olle. *Swedenborg*. Stockholm: Proprius, 1976.

Heinrichs, Michael. *Emanuel Swedenborg in Deutschland*. Frankfurt am Main: Bern Cirencester/U.K. Lans. 1979.

Helander, Hans. *Festivus Applausus in Caroli XII in Pomeraniam suam Adventum*. Upsala: Almqvist & Wiksell, 1985.

Sjödén, Karl-Erik. *Swedenborg en France*. Stockholm: Almqvist & Wiksell, 1985.

INDEX

HENRY SÖDERBERG, AUTHOR

Henry Söderberg was born in Linköping, Sweden in 1916. He graduated with a Degree in Law from the University of Stockholm in 1942.

During World War II, Mr. Söderberg worked for the YMCA International War Prisoners Aid with headquarters in Geneva, Switzerland. His responsibility as field delegate was to visit the prisoner of war camps in Germany, Austria and Poland and to cater to leisure time needs of Allied prisoners, primarily American and British. He continued after peace was declared with the same position for one year, now among retained German prisoners in camps in Belgium, Holland and Luxembourg. During 1946 and '47, Mr. Söderberg traveled extensively on lecture tours throughout the United States.

After the war he became an honorary member of both the American and British officers ex-prisoner of war associations and he continues to participate several times a year in their activities.

In 1947, Mr. Söderberg joined the Swedish Civil Aviation Administration specializing in air political matters. In the 1950's and early '60's he represented the Swedish Government to the United Nations' special agency for civil aviation, ICAO, in Montreal, Canada.

(continued)

Mr. Söderberg is a recipient of The Palms of Order of the Crown, Belgium, 1946; The Medal of Merit, Swedish Red Cross, 1946; His Majesty's Medal for the Cause of Freedom, Great Britain, 1947; and the Order of the North Star, Sweden, 1966.

He joined Scandinavian Airlines System, (SAS) in 1966 as Director of IATA matters. In 1969 he became a Vice President and Chief of Aeropolitical and Foreign Affairs for SAS.

Having been thus actively engaged in aviation through its formative years, it seemed natural that after retirement in 1981, he be called back by SAS to write a history of aviation stressing the roll played by Scandinavians in international aviation.

It was while researching for this project at the Massachusetts Institute of Technology that Mr. Söderberg, quite by accident, came across the first rational aircraft design, produced by Emanuel Swedenborg (as early as 1714). After continuing research on several continents for this book, Mr. Söderberg was surprised and delighted to find the original plans for Swedenborg's airplane preserved in the Diocesan Library in Linköping—the city of his birth.

The recognition of this design being "the first" rational design for heavier-than-air flight by such prestigious institutions as the Royal Aeronautical Society in London, and the Smithsonian Air and Space Museum in Washington, DC led to Mr. Söderberg's further research and writing of this book.